AF572269

# Environmental Pollution and Protection

# Environmental Pollution and Protection
## An Introduction

Edited by
Dhandapani Alagiri
E Naveen Kumar

2012

Icfai Books
The Icfai University Press

ENVIRONMENTAL POLLUTION AND PROTECTION: AN INTRODUCTION

**Editors:** Dhandapani Alagiri and E Naveen Kumar



First Edition: 2012
Printed in India

*Published by*

This book is published by IUP.
University Campus, Agartala-Simna Road,
P.O. Kamalghat Sadar, Agartala – 799210, Tripura (West)
E-mail: info@iupindia.org
Website: www.books.iupindia.org



**ISBN:** 978-81-314-2714-9

# Contents

# Overview

## Introduction

The World's resources are being continuously consumed at an increasing pace and the destruction of the natural resources and environment as a consequence are also increasing at a fast pace.

The environment conservation is one of the hot topics of the day. To understand the gravity of the topic, one needs only to look at the statistics of the last few decades and see the manifold increase in all things consumed and the degradation this has caused. Over the latter half of the last century, in which the modern political and social grouping of nations mostly evolved, the world population rose from 2.5 billion to 6 billion people. The world economy has expanded seven times. Production of food grains has tripled and so has the usage of water. Globally, the number of automobiles grew from 53 million in 1950 to nearly 600 million in 2005. And with the introduction of commercial jet aircraft in the late 1950s, air travel volume ballooned, from about 28 billion passenger-kilometers at mid-century to more than 3 trillion in 2005.

These consumptions by the world population have tremendously increased the consumption of natural resources. As stated earlier, every consumption listed above uses the naturally available resources for its production. Naturally, with the explosive growth in demand, the competition for resources is becoming increasingly intense. This competition is the concern and cause of environmental degradation of the Earth.

This unchecked use of natural resources has led to Wanton environmental degradation and pollution. In recent decades, this pollution of the natural environment has gathered pace and the environmental conditions have worsened further, even though the problem has been taken note of and is being increasingly addressed to even though in a limited way.

The result is very much evident in the following:

**Forests and Land Degradation**

After the start of massive industrialization in·the recent world order, more than half of the world's forest cover has been lost to the industrial developments since the last century and a half. It is estimated that each year around 20 million hectares of the forests are being cleared for development of land for other purposes. This deforestation drives costs the environment very dearly by way of climate change and affecting rains and precipitation across the world. The earth's biological diversity is also being severely affected and this, in turn, puts strain on agriculture practices and livestock in a cyclical way. This also pushes many species of plant and animals into extinction.

**Food Supply**

Food supply lags behind the increase in population in several parts of the globe, especially in the poorer countries. The intensive cultivation practices using mechanized farm equipment and inorganic pesticides have also led to degradation of the arable lands to a considerable extent. The degradation of lands by release of pollutants from several industries has also added to the problem.

The degradation of agricultural lands along with adverse changes in climatic conditions in several parts of the globe, coupled with industrial pollutants have made the task of feeding the world's population a more challenging task. Similarly, potable water has also become a luxury for many populations around the world. Lack of proper drinking water and also sanitation has been a major cause of diseases in the poor countries and, each year, an estimated 15 million people die from contagious diseases and lack of proper medical facilities.

### Air Pollution

Air pollution is another major cause for concern around the globe. The pollutants such as $CO_2$; emitted by the ever increasing vehicular traffic and industries are increasingly polluting the atmosphere and has been the cause of several climatic changes including global warming. The greenhouse gases, as these are called, block the atmosphere and also play an important role in ozone depletion. Another effect of these gases is what is called global dimming. The greenhouse gases effectively block the vital solar energy in parts of the globe and it becomes difficult for vegetation to thrive in these areas.

In major cities that lack pollution controls, the problem of haze leading to health problems, such as eye irritation and respiratory problems are increasing. Cities such as Bangkok, Manila, Nairobi, Mexico City and New Delhi are some of the worst polluted cities.

Air pollution has also been the cause of acid rains in several parts of the globe. When sulphur dioxide and nitrogen oxide combine with water vapour, it results in acid rains. Other pollutants that can combine with the above are the soot and carbon monoxide emissions. Acid rains can cause damage to vegetation and also degrade the soil.

### Coastlines and Oceans

With the massive pollution of the atmosphere, the global temperatures are gradually increasing and this may result in the meltdown of the Arctic and Antarctic ice shelves and may lead to the increase in sea levels. This possibility will put some 60% of the world's population in peril as most of the coastal cities will be in danger of submersion. The fisheries and the marine life are also getting affected as a result of global warming.

Decreased availability of food for the masses and lack of availability of safe drinking water and proper sanitation facilities have led to the miserable living standards for the population under poverty line in many of the African and the Asian countries.

### Others

Many other forms of pollution such as e-waste, which arises from improper disposal of the used electronic products, pollution from usage of heavy metals etc are also increasing day by day. The heavy metals that have been identified to be dangerous to health include lead, mercury, cadmium, arsenic, copper, zinc, and chromium. Such metals, even though are present naturally in the soil in meagre amounts, when become concentrated in particular areas present very serious hazards to health. For example, arsenic and cadmium can cause cancer. Mercury can cause genetic damage, while copper, lead, and mercury can cause damage to the brain. Another serious concern is the disposal of the spent nuclear fuel. This spent fuel emits radiation which is very harmful to humans and environment and can cause death if exposed to higher amounts. There has not been any transparent mechanism in the disposal of the spent nuclear fuel. Another pollution increasing day by day is the noise pollution. High decibel industries and transport modes are causing continuous noise pollution and this may cause severe health problems including ear problems and mental stress to the exposed population. Some other pollutions are visual, light etc, which are increasing with changing lifestyles.

### Environmental Disasters

Some of the major environmental disasters that have happened over the years are[1]:

---

[1] *http://en.wikipedia.org/wiki/List_of_environmental_disasters, http://207.150.180.135/List_of_nuclear_accidents#2000s*

The Dust Bowl in the United States (1934-1939) – 500000 displaced.

Mercury poisoning in Minamata, Japan – Minamata disease (1950s & 1960s) – 2,265 victims had been officially recognized (1,784 of whom had died) and over 10,000 had received financial compensation.

Kyshtym – A serious accident occurred during the winter of 1957-58 near the town of Kyshtym in the Urals. A Russian scientist who first reported the disaster estimated that hundreds died from radiation sickness.

Itai-itai disease, due to cadmium poisoning in Japan – 184 victims have been legally recognized since 1967, of whom 54 were recognized in the period from 1980 to 2000.

Bhopal disaster (1984) in India – 20000 total deaths since the disaster making it the largest industrial disaster.

Chernobyl (1986) – The world's worst nuclear accident occurred after an explosion and fire at the Chernobyl nuclear power plant. It released radiation over much of Europe. Thirty-one people died in the immediate aftermath of the explosion. Hundreds of thousands of residents were moved from the area and a similar number are believed to have suffered from the effects of radiation exposure.

Sandoz chemical spill into the Rhine river (1986) – Wiped out the remaining fishes and aquatic life in the polluted Rhine river.

The severe damage to the Ok Tedi river system in Papua New Guinea by mining operations – The discharge caused widespread and diverse harm, both environmentally and socially, to the 50,000 people who live in the 120 villages downstream of the mine.

Destruction of the old growth forests – around the world destroying ecological balance.

Kursk (2000) – The Russian submarine Kursk sinks in the Barents Sea after an apparent internal torpedo accident, killing 118. Russia eventually recovers the submarine's nuclear reactor and states that the submarine had carried no nuclear weapons.

AZF Explosion at a Toulouse chemical factory (2001) – caused 29 deaths, 2,500 seriously wounded and 8,000 light casualties.

Mihama (2004) – An accident in the nuclear power of Mihama, in the Fukui prefecture 320 km northwest of Tokyo causes five deaths and seven injuries, the deadliest nuclear power plant accident in Japan.

2005 Jilin chemical plant explosions in China – Over 10 000 people evacuated.

The Sydney Tar Ponds and Coke Ovens sites in the city of Sydney, Nova Scotia, Canada-known as the largest toxic waste site in North America.

Release of cyanide, heavy metals and acid into the Alamosa River, Colorado in US from the Summitville mine- caused death of all marine life within a 17 mile radius.

Release of 20,000 gallons of lethal chemicals (metam sodium, tradename Vapam) into the Upper Sacramento River near Dunsmuir, USA – caused the death of all marine life within a 38 mile radius.

Apart from these, there have been major oil spills in the oceans around the world such as the Gulf War oil spill in 1991caused by war or the Atlantic Empress tanker oil spill off Trinidad and Tobago in 1979 also caused enormous damage to marine environment.

**Environmental Protection**

To control the damage caused and save the planet from further damage from pollution and degradation, several steps have been taken over the years. The following is a detailed list of such efforts:

**Major Environmental Events**

1970 – OECD (Organization for Economic Co-operation and Development) Council at ministerial level held. OECD Environment Committee formed.

1972 – United Nations Conference on the Human Environment held in Stockholm. Declaration of The United Nations Conference on The Human Environment adopted. United Nations Environment Program established.

1974 – OECD Council adopted the Declaration on Environmental Policy, which calls for assessment of the environmental impact on major public and private activities.

1987 – Vienna Convention for the Protection of the Ozone Layer Montreal Protocol on Substances that deplete the Ozone Layer adopted.

1988 – Intergovernmental Panel on Climate Change (IPCC) established.

1989 – Basel Convention on the Control of Transboundary Movements of Hazardous Wastes and their Disposal adopted.

1992 – United Nations Conference on Environment and Development (Earth Summit) held in Rio de Janeiro. Earth Charter formulated, Agenda 21 adopted and the United Nations Framework Convention on Climate Change (UNFCCC) adopted (effectuated in March 1994).

1995 – UNFCCC: The First Conference of the Parties (COP1) held in Berlin. IPCC Second Assessment Report published.

1996 – COP2 held in Geneva. ISO14001 launched.

1997 – COP3 held in Kyoto. Greenhouse gas reduction targets for developed countries decided.

1998 – COP4 held in Buenos Aires.

1999 – COP5 held in Bonn.

2000 – COP6 held in the Hague.

2001 – COP6 (Part 2) held in Bonn. Persistent Organic Pollutants (POPs) Treaty.

IPCC Third Assessment Report: Climate Change 2001.

2002 – Emissions trading markets opened in UK and other countries. World Summit on Sustainable Development held in Johannesburg. COP8 held in Delhi.

2003 – EU Directive on the Restriction of the Use of Certain Hazardous Substances in Electrical and Electronic Equipment (RoHS Directive) adopted. COP9 held in Milano.

2004 – COP10 held in Buenos Aires.

2005 – Kyoto Protocol published. COP11 and MOP1 held in Montreal.

2006 – RoHS compliance came into effect. COP12 and MOP2 held in Kenya.

The efforts of the world community towards this end started off really with the summit at Rio de Janeiro, Brazil in June 1992, even though there had been many efforts early including the formation of United Nations Environment Program. The summit was attended by over 150 nations including the United States of America, the treaty signed called Rio Climate Treaty came into force in March 1994. The Rio Climate Treaty set an overall framework for climate protection. But the Rio treaty failed in that it did not set any binding emission limits for nations. There was no voluntary cut down in emissions as envisaged in the treaty, except in Eastern Europe, where the emissions were down only because of the close down of non competitive industries. This brought in the need to have an emissions protocol with binding limitations especially on developed countries.

This was agreed upon in the Kyoto protocol in 1997.

**Kyoto Protocol**

But the Kyoto protocol met with opposition from US right from the beginning. The opposition was that the protocol might threaten US economical interests and the protocol will favour the developing countries. But with he ratification by Japan and by Russia the Kyoto protocol went into effect from Feb 16 2005 onwards. But in its initial stages it is expected to have only a marginal effect on global emissions trends. With the US and Australia keeping out of the protocol and with no effective compliance mechanism even for those who have ratified, Kyoto is not expected to produce great changes.

Apart from the official efforts, several non governmental organizations such as Greenpeace are also working to save the planet by forcing the governments across the globe to be more environment-friendly.

**The Future**

With the slow progress in all the concerned areas, it seems that, unless some drastic cuts in emission limits are agreed upon, there will not be any real reduction in environmental degradation. With the US still not considering the issue serious enough and insisting on bilateral or zonal agreements, instead of participating in global initiatives and other countries even though they have ratified the Kyoto protocol, still only making meager efforts in arresting the environmental degradation, it will take some more time for the world to see stricter agreements in place and a better environment to live in.

In this scenario, when environmental pollution and its protection is an important issue before the world, this book, **Environmental Pollution and Protection: An Introduction** makes a comprehensive study assessing the issues involved and analyzes the important trends and developments that harm or protect the environment. The book presents a series of well-written articles that bring out the essence of the happenings in environmental issues.

The collection of articles in the **First Section**, introduces the reader to the environmental **Introduction**; the **Second Section** of the book concentrates on the **Government Initiatives** and the **Third Section** on **Corporate Responsibilities and Environmental Accounting** in protecting and saving the environment.

The first article "**Environmental Awareness and Pollution Control Strategies Around the World**" by *Dhandapani Alagiri* and *E Naveen Kumar,* aims to bring about the status of environmental problems across the globe as well as in India and the factors responsible for the same. Overexploitation of the natural resources, be it land or water, and the industrialization process have resulted in considerable environmental degradation of these resources, along with the pollution of these and other environmental parameters. The demographics and socio-economic factors have also influenced the environment, though indirectly. Environmental movements arise largely due to the clash between economic development and exploitation of natural resources in the process of doing so. The rise of such movements and the global agreements in this regard are studied. The article also deals with the interventions made by the governments by way of policy reforms, as well as the non-governmental initiatives in minimizing the environmental degradation.

The second article "**The Political Economy of Global Warming and its Changing Strategic Role**" by *Dipankar Dey* focuses on the changing importance of climate change (with special reference to 'global warming'), in a TNC (Transnational Corporation)-dominated world order. The article starts with the genesis of the global warming concept—its scientific basis, economic consequence and political motives. It also covers the major developments in the global warming issue during the period between the Rio Earth Summit and ratification of Kyoto Protocol in 2005, while highlighting the fact that within a decade of its declaration, the basic objectives of the Rio Earth Summit have been distorted systematically, confirming the suspicion that from its inception, the objectives were not taken seriously by the leaders. The article concludes with the observation that in an integrated global economy, where TNCs have developed alliances with their local partners for further consolidation of their power, an international alliance of consumer and civil societies to safeguard common citizens' interest across the globe is essential.

*M Ramanjaneyulu* in the next article "**Environmental Threat for Developing Economies**", discusses the environmental devastation which is taking place by energy consumption and the carbon dioxide emission which is a potential pollutant for degradation of the environment. The mad rush for rapid economic growth has resulted in degrading the environment. Statistics show that the developing economies are all set to take over the developed countries in causing pollution and disturbing the ecological balance by emitting toxic gases like carbon dioxide ($CO_2$). High energy consumption by the developing countries, combined with population explosion and growing needs, is the reason behind the emission of $CO_2$, which has affected the environment. The author also analyzes various pollutants and their impact on the global environment and explores the global environmental scenario with special reference to India and its National Environmental Policy, against the backdrop of the Global Environmental Facility Program.

The fourth article, "**Environmental Crime in Global Context: Exploring the Theoretical and Empirical Complexities**" by *Rob White* deals with concerns across a wide range of environments (e.g., land, air, water) and issues (e.g., fishing, pollution, toxic waste) involving conceptual analysis as well as practical intervention on many fronts, and includes multi-disciplinary strategic assessment (e.g., economic, legal, social and ecological evaluations). It also features the undertaking of organizational analysis, as well as investigation of 'best practice' methods of monitoring, assessment, enforcement and education regarding environmental protection and regulation. The major issues relating to environmental crime are the trans-border movement and dumping of waste products, the illegal traffic in real or purported radioactive or nuclear substances and the illegal traffic in species of wild flora and fauna. The paper discusses issues relating to scale (e.g., local, regional, global) and impacts (e.g., immediate, long term), as these relate to matters pertaining to definition and jurisdiction (e.g., international law and law enforcement, eco-human rights). In addition to mapping out complexities involved in examination of environmental harm, the paper attempts to identify potential directions for further theoretical reflection and empirical research.

The first article in the second section "**National Environmental Policy 2006: Some Observations**" by *Usha K R,* describes the salient features of the National Environment Policy (NEP) 2006 released by the Ministry of Environment and Forests. It highlights the imperative need to maintain the balance between economic development and environmental protection and focuses on the key environmental challenges that the country faces relating to the nexus of environmental degradation with poverty in its many dimensions, and economic growth. Research on feasibility of social forestry, provision of safe drinking water, strict enforcement of the rules relating to the water, air and sound pollutions are suggested as timely measures to safeguard the environment from certain threats. The dominant theme of this policy is that, while conservation of environmental resources is necessary to secure livelihoods and well-being of all, the most secure basis for conservation is to ensure that people dependent on particular resources obtain better livelihoods from the fact of conservation, than from degradation of the resource.

The second article "**Environmental Governance in India**" by *Geetanjoy Sahu* studies the evolution of the concept of environmental governance in India. It discusses in detail the legal framework and constitutional provisions for environmental protection. Integration between Environmental Policy and Economic Planning are discussed. In the initial period, India did not give much importance to environment. However, since 1970s, there has been an increasing concern for environmental protection and improvement in Indian policy and planning process. It was in the Fourth Five-Year Plan (1969-74) that for the first time environmental issues were given high priority. The paper further highlights the role of judiciary in environmental governance and briefs contributing factors and emerging issues in the environmental governance process .It also explores the significance and implications of these issues to ensure a healthy environment and aims to protect and improve the environment.

*Krishnayan Sen* in the succeeding article **"A Public Health Approach to Environmental Protection in India"**, starts with addressing environmental rights affecting public health endeavor in the international context. After analyzing the international treaties and conventions relating to environmental and public health protection, the author seeks to locate the right to health as a norm in customary international law. A public health approach to environmental protection potentially touches upon all spheres of human activity, and claims to override or trump other considerations. It also examines the emerging trends in the application of environmental rights in domestic law, enunciating the position of the right to health and environment under the Indian Constitution and submits as to how the enjoyment of one is intrinsically linked with that of the other. It would be observed that, while much of the current legal debate occurs at the international level, the detailed experience of environmental rights in domestic jurisdictions (India being the case in point), offers rich instruction, simultaneously showing promising ways to move forward as well as examples of mistakes to be avoided. Equal emphasis on international and domestic law is, therefore, deliberate.

The next article **"Global Warming and Clean Development Mechanism Projects: State and Trends in India"** by *Jatinder S Bhatia* and *Harsh Bhargava,* enunciates that Greenhouse Gases (GHGs) are a major source of global warming and are responsible for the depletion of the ozone layer. As a first major step to bring down the GHG concentration in the atmosphere, the Kyoto Protocol came with itself a vast pool of Clean Development Mechanism

(CDM) projects, which are currently under operation in many countries. The CDM was designed to assist developing countries in achieving sustainable development by promoting environment-friendly investment from industrialized country governments and businesses. India acceded to the Kyoto Protocol in August 2002 to fulfill the prerequisites for implementation of CDM projects, and is fast catching up with other countries to deliver a diverse portfolio of projects. India is on the top of the list in terms of number of ongoing projects, in different host countries throughout the world. As on April 1, 2006, out of a total of 693 projects worldwide, 267 were being carried out in India. India offers tremendous CDM opportunities in the area of renewable energy, land-use, land-use change, and forestry, agriculture and livestock, Power, Industrial energy and Transport. One of the most challenging aspects of building a national CDM strategy is to ensure an active support from all sectors of society (civil, NGOs, private and public sector) and different sectors of the economy (industry, energy, agriculture, forestry).

The fifth article **"The Socio-Technical Challenges of Safe Water Supply in Rural Bangladesh"** by *Wahidul K Biswas* and *John Merson*, suggests that the shortage of potable water in rural Bangladesh is the result of lack of water conservation strategies, inaccessibility to affordable water supply options, and the weak institutional framework The rural poor in the inland and coastal zones suffer from a scarcity of potable water due to the presence of arsenic and salinity in the groundwater. In the arsenic- and salinity-affected areas of Bangladesh, rural people still struggle to procure safe drinking water due to technological, institutional and policy barriers. This paper identifies and analyzes these barriers by using a policy orientation approach. Social and decision processes are mapped to demonstrate the historic and present-day trends and conditions for water supply. Options for different potable water technologies are assessed and alternative future development scenarios using models of different water supply programs are projected for the provision of safe water for the rural communities in Bangladesh. The alternative scenarios describe a self-sustaining mechanism which will enable the poor people in rural areas to have easy access to safe water, which will further help them to secure ownership of the capital-intensive water purification technology.

The next auricle **"US-International Climate Change Approach: A Clean Technology Solution"** by *Paula Dobriansky* features the US Administration's international engagement on climate change issues centering on five key ideas, all of which extend from and build on its own experience here in the United States. A successful international response to climate change requires developing country participation, which includes both near-term efforts to slow the growth in emissions and longer-term efforts to build capacity for future cooperation actions, absent the participation of all major emitters, including developing countries, the goal of stabilizing GHG concentrations will remain elusive. The US has initiated and is participating in a range of new technology initiatives designed to meet climate and clean development goals. The article briefly highlights a few of the most significant partnerships like Group on Earth Observations, International Energy Research and Development Partnerships, The Methane to Markets Partnership and World Summit on Sustainable Development Partnerships. Meeting the challenge of the expected future growth in global energy demand and reducing greenhouse gas emissions will require a transformation in the way the world produces and consumes energy over the next generation and beyond. It will require new ways of collaborating with US partners to break through long-standing stalemates. This is why US is leading global efforts to develop and deploy breakthrough technologies for both the developed and developing world.

The first article in the third section **"Corporate Social Responsibility"** by *Paul Watchman* and *Malcolm Forster,* emphasizes that CSR is part of an international drive towards transparency and accountability of business activities and a way of monitoring how businesses perform against environmental, ethical and social indices. Rapid development of company and industrial activities, without proper precautionary measures for protection of environment, are causing pollution. Hence, it has become a necessity for a company to comply with the regulatory norms for prevention and control of pollution to the extent that the regulatory norms are to be complied with by adopting clean technologies and improved management practices. Corporate Social Responsibility is one of the responses to the imbalances resulting from the acceleration of the globalization process. The company aims to protect and restore the environment by minimizing the use of resources and energy, by decreasing waste and harmful emissions, from its activities.

The next article "**Extended Producer Responsibility for Environmental Protection: The Concepts, Relevance and Lessons**" by *C S Shylajan* and *G Radha Krishna,* discusses the concept of Extended Producer Responsibility (EPR) which has emerged as a policy tool for managing the environment by promoting recycling, and thus reducing waste. The concept implies that responsibilities, which were traditionally assigned to consumers and waste management authorities, are to be extended to the manufacturer of the products based on the 'producer pays' principle. It provides incentives to producers for making Design for Environment (DfE) changes to products that would reduce waste management costs. This paper reviews important EPR policy tools implemented in countries around the world and discovers that many industrialized countries have implemented EPR policies and have been successful in achieving the targets of reducing waste. Some important principles which are relevant for implementing EPR programs effectively are also discussed.

*Arup Choudhuri* in the succeeding article "**Measuring Sustainability and Green Reporting: The Role of Environmental Audit**", emphasizes the role of environmental audit in measuring sustainability and green reporting. It is apparent that the area of environmental auditing is probably one of the most dynamic and important subjects to come to the attention of the corporate world. It is an area with which internal and external auditors, and engineers as well as accountants must become familiar to make their companies survive in the contemporary world of competition. By using environmental accounting as a strategic weapon, companies can prosper by acquiring brand equity and goodwill. To keep an efficient environmental accounting system in place, the two major areas to be considered are valuation vis-à-vis quantification of environmental issues and control through proper environmental auditing. This article discusses environmental audit techniques and strategies, various categories of environmental audits like compliance audit, Environmental Management System Audits, Pollution Prevention Audits etc. and management's ecological responsiveness.

The next article "**Beyond Environmental Accounting: Triple Bottom Line Approach**" by *Bibhuti Pradhan* and *Sanjib Pattnaik*, explores the advancement of corporate responsibility concepts and Triple Bottom Line (TBL – Economic, Social and Environmental sustainability) as an approach in achieving corporate sustainability. It also presents the path to sustainability that firms look for and work out strategies to guarantee financial success, while, at the same time, managing its environmental and social responsibilities. The notion of sustainability has grown to encompass economic, environmental and social dimensions. But today, the primary concern of many organizational managers is how to bring together these fundamental, yet seemingly disparate pillars of sustainability to form the TBL of economic, environmental and social performance. Ultimately, this article seeks to find a new set of management strategies and tools through TBL approach for managing and balancing corporate responsibility.

The final article "**Red Carpet for One-Stop-Shop Environmental Courts in India**" by *Lakshmi Lella* analyzes the pros and cons of constituting separate environmental courts in India, and advocates for the same. It proposes to look into the need, circumstances, judicial opinions, and constitutional mandates for constitution of such special courts; despite the fact that there exist environmental benches and tribunals. The multifaceted legal issues relating to science and technology are increasing day by day, leading to complex environmental litigations. The courts thus stand in a delicate position, as they have to choose between environment and economic development, while both are equally important for the development of the nation. It further analyzes the 186th Report of Justice M Jagannadha Rao, Law Commission of India, on the 'proposal to constitute separate environmental courts.

# SECTION I

# INTRODUCTION

# 1

# Environmental Awareness and Pollution Control Strategies Around the World

*Dhandapani Alagiri and E Naveen Kumar*

***This article aims to bring about the status of environmental problems across the globe as well as in India and the factors responsible for the same. Overexploitation of the natural resources, be it land or water, and the industrialization process has resulted in considerable environmental degradation of these resources along with the pollution of these and the other environmental parameters. The demographics and socio-economic factors have also influenced the environment, though indirectly. Environmental movements arise largely due to the clash between economic development and exploitation of natural resources in the process of doing so. The rise of such movements and the global agreements in this regard are studied. The article also deals with the interventions made by the governments by way of policy reforms, as well as the non-governmental initiatives in minimizing the environmental degradation.***

## Introduction

Environmental pollution and degradation has been the bane of the industrialized world. Over the last two centuries, there has risen a world that has been more and more driven by machines and materialistic interests that has led to a fast evolving and changing lifestyle. This, in turn, has made the world population to rapidly use the naturally available resources and convert them into products that are increasingly non-recyclable and polluting to the natural environment. This process of fast paced industrialization and more materialistic lifestyle has been happening unabated for the last two centuries and only now, there has been some realization to the damage caused by way of polluting the environment. There have also been many struggles from time to time against the indiscriminate use of natural resources and pollution of the environment. But these struggles have taken a more solid form only in the recent decades. The governments around the world have also woken up to the enormity of the problem and several non governmental organizations (NGOs) and corporate bodies have also taken up the issue and are working to create environmental awareness and devising pollution control strategies that could reduce the enormity and severity of the problem of environmental pollution and degradation.

## Industrialization

The industrialization process that started in England in the late eighteenth and early nineteenth century can be taken to be the beginning of the pollution problems of the present era even though environmental pollution has been as early as human civilization. Before the industrialization process set in, in Europe, the world civilization had been more sensibly using the resources available to it and there had not been any extravagant or fast paced devouring of natural resources and resulting complications. The previous world powers that existed before the advent of industrial Europe had been more adaptive of the facilities/lifestyles that had been less changing over several centuries and, hence, less harsh on the natural resources. But with the advent of the industrial age, all the countries of the world started using the resources indiscriminately and that too in a much faster pace and this has led to more and more environmental pollution and degradation. This fast production and consumption for over two centuries have put enormous strain on the resources and the environment.

## Changing Lifestyles

From the last decades of the last century, there has been a new revolution driven by the innovation in the information technology and a more mobile lifestyle. The new revolution is led by the increased globalization, increased industrial activity and technological innovations. Globalization has led to expansion on businesses across the borders and increased production. The revolution in telecommunication and information technology has also added to this fast paced growth across the globe. The power of large corporate houses has also increased in this period. All these have contributed to a fast paced lifestyle that generates far more final products than needed thus leading to a general waste of natural resources and also causing pollution and degradation.

If this fast paced economic growth, without any major concerns for the environmental problems being created goes on unchecked, which is most likely, then there will be a continued degradation of the environment which might lead to several conflict situations across the globe. The countries that bear the brunt of the environmental degradation and climate change will see large scale migration and increased poverty which will lead to tensions between countries. The large scale environmental degradation has already led to major illnesses across the globe and will in future lead to more contagious diseases.

## Land

With the advent of agricultural practices increasing and the shrinkage of the forest cover, the efforts to maintain forests is a major challenge in the developing world. However, in the developed countries, where the population growth has stabilized, the major awareness concerns relate to forest and biodiversity conservation. While the developing countries are taking up the conservation of biodiversity, the major biospheres that are important to sustain the global climate and environment and are in need of protection fall in the Latin American, African and Asian continents. Lack of proper co ordination between countries and the general lack of awareness and apathy by poor countries and developing countries have led to habitat loss for wild life and loss of forest cover all over the world. The developing countries blame the developed countries for the gross environmental violations of the past and expect them to increasingly share the burden of saving the environment worldwide.

## Water

Many of the regions of the world are increasingly facing the problem of polluted water supply, be it of underground or surface sources. Some of the areas of Africa, especially in the drought prone East, face severe problem and many people die regularly for want of adequate safe drinking water as well as water borne diseases. Water borne diseases form the single largest cause of illness and resultant death worldwide. Some 1.7 billion people, more than one-third of the world's population, are without safe water supply. In addition, an estimated one-quarter of the world population is suffering from chronic water shortages in the beginning of this century[1]. Whereas in the poorer countries, efficient management of the available water resources and supply of safe potable water is the main concern, the developed countries concentrate more on protecting the water resources and reclaiming the polluted water bodies by detoxifying them. Formation of mega cities (see box below) with their bursting populations is another area of increasing awareness and concern, these cities indiscriminately suck the under-ground water and this results in land degradation and seepage of sea water in the coastal cities. These cities produce huge amount of sewage and other industrial waste products that degrade the land and water resources of the surrounding areas from where these cities inevitably have to source their water needs. The widespread need for water has also led to tensions between riparian countries sharing river basins on the basis of waters, pollution and strategic use.

| Rank | City/Urban area | Country | Population in 2006 (millions) |
|---|---|---|---|
| 1 | Tokyo | Japan | 35.53 |
| 2 | Mexico City | Mexico | 19.24 |
| 3 | Mumbai (Bombay) | India | 18.84 |
| 4 | New York | USA | 18.65 |
| 5 | São Paulo | Brazil | 18.61 |
| 6 | Delhi | India | 16.00 |
| 7 | Calcutta | India | 14.57 |
| 8 | Jakarta | Indonesia | 13.67 |
| 9 | Buenos Aires | Argentina | 13.52 |
| 10 | Dhaka | Bangladesh | 13.09 |

*Source: http://www.citymayors.com/features/urban_areas.html*

[1] *http://www-cger.nies.go.jp/geo1/index.htm*

## Air

The other major concern that plagues the entire world irrespective of developed or developing regions is the air pollution. All the major cities of the world face severe air quality problems. Apart from the poor quality of air over the cities, there are also problems of air pollution over vast country sides too. Acid rain caused by industrial pollutants can play havoc over vast areas apart from their place of origin. Such problems also turn trans-country and transcontinental leading to air pollution in other countries as well. The smog witnessed in Malaysia by the forest fires of Indonesia is one such example (see box below).

**Smog Alert in South-east Asia**

Thick, acidic smoke from rainforest clearances in Indonesia is seriously affecting Singapore and substantial parts of Malaysia, including the capital, Kuala Lumpur.

The health authorities are describing the atmosphere in the southern and western parts of Peninsular Malaysia as "unhealthy". Problems with smoke are even worse on the island of Borneo: the Foreign Office warns that "Sarawak has been particularly badly affected, with air quality in some areas reaching 'very unhealthy' levels".

The smoke is from fires raging in the Indonesian island of Sumatra, and the region of Kalimantan in Borneo. The blazes are started deliberately in order to clear rainforest for palm-oil plantations. Such clearances have been taking place during the dry season for the past 10 years.

Meteorologists doubt that the prolonged pall will start to clear until late November, when the northeast monsoon should douse the fires.

*Source: Independent, The (London), November 4, 2006 by Anthony Lambert.*

The damage to the Ozone layer by the industrial and automobile emissions and other green house gases has been continuing without any let off. NASA records that from September 21-30, 2006, the average area of the ozone hole was the largest ever observed, at 10.6 million square miles (27.5 million square kilometres). The blue and purple colors (appears very dark in monochrome) are where there is the least ozone, and the greens, and yellows are where there is more ozone[2].

**Ozone Hole over the Antarctic**

[2] http://www.nasa.gov/vision/earth/lookingatearth/ozone_record.html

Rapid increase in the population and vehicles are the result of fast paced growth of the economies of developing countries especially of Asia. This has led to an enormous increase in demand for energy which is mostly generated from polluting resources and using environment degrading technologies as the fossil fuels – coal oil and natural gas are going to be the main sources of energy.

Another consequence of the global warming due to the air pollution and depletion of ozone layer is the possible raise in sea levels across the globe as the ice sheets over the Arctic and the Antarctic poles continue to melt in increasing amounts. As more than 60% of the world's population live near the coastal areas or dependent on the coast for their livelihood. The coastal areas might come under threat of submergence and this might lead to catastrophic results. The coastal areas are also under threat from various intensive aquaculture practices. Over exploitation of coastal fisheries and development of tourism infrastructure along the coast also affect the marine eco systems along the coast.

## Noise

Another form of pollution that is increasingly causing annoyance and also health hazards is noise pollution. The source of most noise pollution is the motor vehicles and the modern gadgets that produce varying levels of noise. High noise can cause hearing problems, cardio vascular problems and stress. Poor urban planning such as housing areas situated close to industrial areas may also lead to noise pollution.

Apart from the above, there are many other forms of environmental degradation taking place such as e-waste that arises from the improper disposal of the electronic products, accumulation of radio active waste and possible radio active spills etc.

The problems of environmental pollution and degradation are made more complicated by the rapid industrialization and urbanization and the widening gulf between the rich and the poor.

## Environmental Protection Efforts

There have been several movements emphasizing the need to protect the environment over the years. Some of the indigenous communities in India such as the Bishnois are very particular and take it up as part of their religion to protect their surrounding environment.

Various indigenous traditions in India and around the world have stressed on the environmental ethics. These were especially based on developing sustainable civilization by avoiding additional stress on the environment.

Generally Hindu, Buddhist and Jain saints and monks have always sought a natural and peaceful environment for the uplift of the soul and for this retreated to forests. Hindu ethos has always promoted sacred groves associated with various temples and houses. But of late, these sacred groves and temple tanks have been vandalized because of political pressures and misuse of temple precincts and properties by narrow-minded people.

## In Modern Times

There have been many movements in India and around the globe in protecting the environment.

Modern environmentalist movements started somewhere in the late nineteenth century in the USA as efforts by individuals. The formation of the Yosemite national park can be attributed to the tireless efforts of Henry David Thoreau, a noted conservationist of his times.

The dawning of the twentieth century saw the emergence of conservation and environmental protection in a big way. The endangering of the American bison and some other species kick started the environmental movement in a big way. The National Park Service was formed as a result in the year 1916.

But these efforts gathered momentum only in the 1950s and 1960s when the effects of DDT on public health came to analyzed and known for the general public. From this time onwards, public sympathy for environmental causes was aroused and this has given more weight for the efforts. More and more voices were raised against environmental degradation and also against more forms of pollution such as noise, oil spills, e-waste etc. Many noted environmental protection groups such as Greenpeace came into existence to champion the cause.

In the 1970s, India had the *Chipko* movement against the indiscriminate cutting down of the trees. The 1970s and subsequent years saw more environmental movements such as the *Back to the Land* which inspired people to move back to the rural areas. *CITES* (the Convention on International Trade in Endangered Species of Wild Fauna and Flora) an international agreement formed as a result of the resolution adopted in the year 1963 at a meeting of members of the World Conservation Union (IUCN) also helped in protecting endangered wildlife and, hence, environment. The United Nations is also actively pursuing environment protection starting from the formation of UNEP in the early 1970s. (see box below)

**UN: Global Warning on Par with War**

UN Secretary General Ban Ki Moon made a dramatic call to preserve the environment Thursday and warned that climatic changes and global warming, "are as dangerous for humankind as wars."

In a speech to open a conference on environment, Ban Ki Moon promised he would take advantage of the first meeting of the G-8 Group in Germany to highlight the urgency of this topic. Before an audience of UN University students, Ban Ki said the priority of the United Nations is still focused on prevention and conflict resolution.

"But the danger that war represents the mankind and our planet is comparable with the climatic crisis and global warming," the UN Secretary General declared.

A panel of experts recommended Wednesday that the UN should be ready to give humanitarian help to tens of millions of "environmental refugees" in the world.

"In the next decades, changes in environment because of the combinations of droughts, floods and loss of cultivable lands will be the main causes for wars and conflicts," he said. In his two first months as Secretary General, Ban Ki Moon has reiterated that climatic change has priority in his agenda.

United Nations, Mar 1 (Prensa Latina)

*Source: http://www.plenglish.com/article.asp*

## Governmental Initiatives

There seems to be a dearth of consensus among the world governments and other international bodies on the future course of action. The lack of consensus on the part of major powers and opinion leaders affect coordination of the efforts (an example in this regard is the refusal of USA and Australia to ratify the Kyoto Protocol).

Nevertheless progress is being made. Standards such as ISO 14000 are increasingly influencing corporate actions and more and more environmental activists are stepping up the drive against pollution and degradation. A detailed list of multilateral agreements on environmental protection under the auspices of United Nations is given in the Appendix.

Major stakeholders across the globe are adopting various strategies that serve to mitigate the degradation, some of which are given below.

## Control Strategies

### That May be Adopted by Governments

a) Regulatory Policy Initiatives that control pollution by having:
   - Emission charges
   - Emission standards

b) Incentive based Policy initiatives that control pollution by:
   - Banking/trading emissions
   - Prohibitively priced limited pollution rights
   - Emission offsets

### That May be Adopted at a Global Level

a) Preparing and communicating a list of national inventory of greenhouse gas emissions using comparable methodologies between nations

b) Developing programs to control effects of greenhouse gases and measures of adaptation to climate change and communicating the same

c) Cooperation on technology related to greenhouse gas emissions

d) Development and management of greenhouse gas sinks and reservoirs (such as reserve forests)

e) Cooperation in adapting to the impacts of climate change

f) Research to enhance scientific knowledge on climate change, the effects of it and the effectiveness of responses to it

g) Exchange of technology developed to counter pollution

h) Emission targets for individual nations that include major greenhouse gases: carbondioxide, methane, nitrous oxide, and synthetic substitutes for ozone-depleting CFCs.

*(Adapted from globalchange.umich.edu[3])*

## Conclusion

There seems to be many hidden problems on account of environmental degradation that are slowly unraveling. One such phenomenon is the un-seasonal rains or the unnatural cyclones as witnessed in 2004 off the coast of South America (see appendix 2). The consensus is that environmental pollution can no longer be taken lightly and efforts to contain the damage need to much more strengthened. Concentrated and coordinated efforts in this direction are the immediate necessity for the sake of future generations of humankind.

*(Dhandapani Alagiri is a Faculty Member at Icfai Business School Research Centre, Chennai. He can be reached adpani1998@gmail.com*

*E Naveen Kumar is a Faculty Associate at Icfai Business School Research Centre, Chennai. He can be reached enkumarssn@gmail.com)*

[3] *http://www.globalchange.umich.edu/globalchange2/current/lectures/pollution_control/pollution_control.html#II*

# APPENDIX 1

## Multilateral Environmental Agreements Currently Under Practice

1. Convention on Biological Diversity (CBD)
2. Convention on International Trade in Endangered Species of Wild Fauna and Flora (CITES)
3. Convention on the Conservation of Migratory Species of Wild Animals (CMS)
4. Convention on Wetlands of International Importance especially as Waterfowl Habitat (Ramsar)
5. United Nations Convention to Combat Desertification in those Countries Experiencing Serious Drought and/or Desertification, particularly in Africa (UNCCD)
6. United Nations Framework Convention on Climate Change (UNCCC)
7. Vienna Convention for the Protection of the Ozone Layer (Ozone)
8. Basel Convention on the Control of Transboundary Movements of Hazardous Wastes and their Disposal (Basel)
9. Rotterdam Convention on the Prior Informed Consent Procedure for Certain Hazardous Chemicals and Pesticides in International Trade (Rotterdam)
10. Stockholm Convention On Persistent Organic Pollutants (Stockholm)
11. World Heritage Convention
12. United Nations Convention on the Law of the Sea (UNCLOS)
13. Agreement for the Implementation of the Provisions of the United Nations Convention on the Law of the Sea (UNCLOS) relating to the Conservation and Management of Straddling Fish Stocks and Highly Migratory Fish Stocks (UNFSA)
14. Convention on the Prevention of Marine Pollution by Dumping of Wastes and Other Matter (London Convention) Pending

15. Cartagena Convention for the Protection and Development of the Marine Environment of the Wider Caribbean Region (Cartagena)
16. Convention for the Protection of the Mediterranean Sea Against Pollution (Barcelona Convention)
17. Convention on the Conservation of Antarctic Marine Living Resources (CCAMLR)
18. Convention on Long-range Transboundary Air Pollution (ECE-LRTAP)
19. Convention on Access to Information, Public Participation in Decision-making and Access to Justice in Environmental Matters (ECE-Aarhus)
20. Environmental Impact Assessment in a Transboundary Context (ECE-EIA)
21. Convention on the Transboundary Effects of Industrial Accidents (ECE-TEAI)
22. Convention on the Protection and Use of Transboundary Watercourses and International Lakes (ECE-Water)

*Source: http://www.un.org/ga/president/60/summitfollowup/060612d.pdf*

# APPENDIX 2

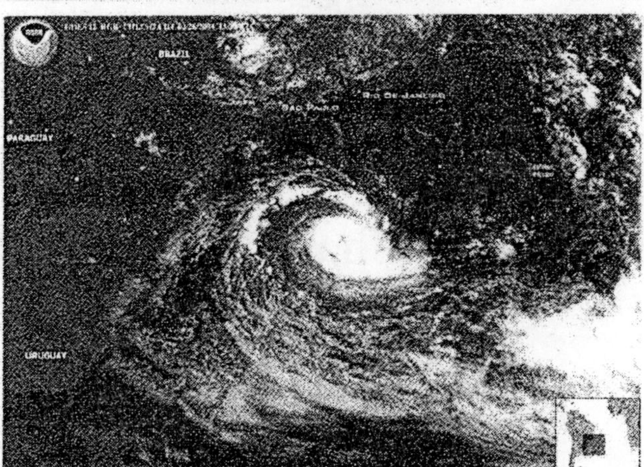


The first recorded South Atlantic hurricane, "Catarina", which hit Brazil in 2004, an indication of climate change?

*Source: NASA.*

# 2

# The Political Economy of Global Warming and its Changing Strategic Role*

*Dipankar Dey*

***To a military strategist of the last century when nation states were eager to retain their sovereign power, weather was a potential weapon. 'Weather' which had a strategic importance as 'weapon' to a nation state during cold war period, is likely to have a strategic importance of different type (as 'economic good') to a TNC dominated world order of this century. Now the emphasis is more on economics than politics. The paradigm shift is very clear and distinct. The study concludes with the observation that in an integrated global economy, where TNCs have developed alliances with their local partners for further consolidation of their power, an international alliance of consumer and civil societies to safeguard common citizens' interest across the globe is essential.***

***This paper has been divided into two sections where the changing importance of climate in a TNC (transnational corporation) dominated world order has been analyzed. The first section, briefly discusses the genesis of the global***

* The 1st section of this study was done by the author in 1992. It was published in The *Frontier*, July 25, 1992, Kolkata, India under the title: 'Of Politics and Global Warming'. The author acknowledges his gratitude to the Editor, *Frontier* weekly for his kind support. In Section I of this updated version, we have reproduced the same study, with few minor editing.

*Source: The Icfai Journal of Environmental Economics, 2006. This article is Published in The Icfai Journal of Environmental Economics, November, 2006; Uploaded as working paper in SSRN on 29th March, 2006, http://hq.ssrn.com/. This is a slightly updated (information on global dimming has been included here) version of the same.*

***warming concept – its scientific basis, economic consequence and political motives. The second section, deals with the major developments in the global warming issues during the period between the Rio Earth Summit and ratification of Kyoto protocol in 2005. This section also highlights the fact that within a decade of its declaration, the basic objectives of the Rio Earth Summit have been distorted systematically.***

## Section I

### Global Warming: The Concept

Till 1970s, the scientists discussed the possibility of a return of Ice-Age. Then from the mid eighties onwards, a section of scientists and policy makers directed our attention to a gradual rise of earth's average surface temperature since late 19th century. Highly technical terminologies like 'green house effect', 'ozone depletion', and 'sustainable development', 'global warming' etc. entered into our vocabulary.

The concern for global warming had arisen from the prediction of a general circulation model (GCM), which indicated a gradual warming of atmospheric temperature resulting in dramatic climate change, melting of polar ice and drowning of low coastal areas in the near future. This complex model, however, provoked a lot of controversy among scientific communities about its accuracy. Scientists disagreed on fundamentals. Questions were raised about the veracity of the claims of the global warming phenomena. Was it a part of a self-correcting natural cycle with no direct association with energy consumption?

### The Cause

The global climate change issue had two components (i) depletion of ozone layer; (ii) rise of atmospheric temperature.

The chlorofluorocarbon (CFC), the man-made 'miracle' gases of 1930s, was identified as the main cause of ozone depletion. There was no controversy on phasing out of this man-made ozone depleting substance (ODS). And to implement the phase out plan, most of the nations unanimously agreed to a protocol commonly known as Montreal Protocol. It was concluded in 1987.

Higher concentration of few inorganic and organic compounds like carbon-di-oxide, methane etc. in the atmosphere have been identified as the major cause behind temperature change. It was claimed that carbon–di-oxide and methane were primarily responsible for global warming and 'green house effect'. The source of carbon-di-oxide and organic gases were mainly fossil fuels like petroleum and coal, decomposed organic materials (biogas) and firewood.

## The Controversy

In March 1992, Dr. S B Isdo, a research physicist with water conservation laboratory, Phoenix, Arizona, USA, had claimed that 'Carbon-di-oxide can revitalize the planet'. Based on his research findings, he argued[1]:

- Carbon-di-oxide was a tremendously effective aerial fertilizer and antitranspirant which stimulated plant growth and development while reducing evaporation of water loss. Due to efficient utilization of water, plant would thrive where previously they had been excluded for lack of this vital resource. With every passing year, more organic matter would be accumulated thereby increasing the level of microbiological activity.
- The biological modulation of climate (not properly considered in the general circulation models – which mainly dealt with physical and chemical process) began with an initial impetus for warming (provided by an increasing carbon-di-oxide content of the atmosphere). According to him, due to warming, phyto planktons grew faster and photo-synthesized at a quicker rate to produce greater quantity of dimethylsuphoniopropionate, a portion of which decomposed to form a compound known as dimethylsulphide that diffused into the atmosphere, got oxidized and converted into sulphuric and methanesulphonic acid particles, which then functioned as cloud condensation nuclei. Some of these nuclei, produced new clouds while other increased the droplet number concentrations of pre-existing clouds, making them more highly reflective of incoming solar radiation. These

effects tended to cool the planet dramatically acting as a break on the initial impetus for warming.

Dr. Isdo claimed that the negative feedback generated out of the above process was about the same strength as the primary 'greenhouse effect'.

It seems that the GCM of global warming was not a scientifically proved and accepted model. Strong arguments in defense of carbon were extended by a section of scientists. And most importantly, it was also told that the initial warm phase of the climate would be temporary which would become normal through a natural corrective process. However, the campaign against carbon continued unabated.

## The Consequences

Once carbon emission and global warming were widely accepted as major threats, the world economy was left with two options:

- Reduce the total energy consumption, drastically
- Reduce the consumption of fossil fuel, biogas, fire wood and substitute these energy sources with a cleaner form of energy.

The first option was not feasible immediately with the existing structure of the world economy. The solution to the global warming problem was linked to the second alternative – i.e., reduction in the consumption of fossil fuel, biogas and fire wood. Vigorous search for a benign and stable alternative source was initiated all over the world.

Alternative energy sources to organic fuels:

- *Non-conventional* sources such as solar, wind etc. Probably the most suitable option but these sources were still in the developing stages and were unable to meet the massive energy requirements of the global economy within a short period.
- *Mega hydel* projects. Projects like Narmada and Aswan Dams had already earned criticism from different quarters for destroying environmental balance with far reaching adverse consequences on the eco-system and biodiversity.
- *Nuclear energy.* Criticized mainly for high risk of health hazards, environmental pollution and high unit cost of production. After the

Chernobyl nuclear disaster in 1986, criticisms against nuclear energy got new momentum worldwide. Proponents of nuclear energy, however, argued that nuclear pollution was less wide spread (local) than the global pollution of organic fuel. They also argued that social cost of nuclear energy would be lower if the cleaning cost of fossil fuel was considered.

The criticisms, which were labeled against nuclear energy, were then redirected against fossils fuel. Public opinions were built systematically to consolidate the position of nuclear energy in the context of global warming. The much-publicized Rio summit which passionately raised the global warming issue had actually strengthened the hands of nuclear lobby. The media and opinion makers echoed the arguments put forward by the nuclear establishment. For example, in India, Raj Chengappa a renowned columnist had advocated, in a prestigious political magazine[2] of India, for considering nuclear energy in a 'big way'. Simultaneously, the Communist Party of India (Marxists) had also demanded a nuclear power plant in West Bengal – the state they were in power since 1977.

## Nuclear Energy: The Gainer

Hence, at the end of the day, the 'global warming' thesis strengthened the position of nuclear energy. The question then is who stands to 'gain' when the balance tilts in its favor?

In the early 1990s, OPEC possessed 70% of global oil reserves and its supplies were forecast to last for another ninety years, whereas for the rest of the world, oil had a life expectancy of only seventeen years[3]. To the developed nations, especially Japan and Western European countries, the crisis was at their doorstep due to near exhaustion of crude reserves. USA and Russia were two exceptions with adequate hydrocarbon reserves. So these two countries were not concerned with the imminent energy crisis faced by their affluent friends. This could be one of the reasons why at Rio-Earth Summit, USA was a 'reluctant participant' and Russian president did not even visit the summit.

Since 1970s, developed countries (and their TNCs), lost their oligopolistic control on oil. OPEC power had changed the energy equations of the world. Till the late eighties, developed countries could not counter OPEC power mainly for two reasons: (i) major accidents in the nuclear plants – one in 1979 at the Three Mile Island (USA) and the other in 1986 at Chernobyl (USSR) halted the growth of nuclear energy which was considered by a section of energy policy experts a viable alternative to fossil fuel, (ii) The Soviet Union with its huge oil reserve was not prepared to align with non communist developed countries – most of which were NATO members.

In the mid-eighties, there was a glut in the world oil market and the price of crude was very low. After the gulf war, OPEC had consolidated its control over production and supply. Crude price was expected to reach to a record high of $21/barrel in 1992. To scuttle this move, European Commission was considering imposing carbon tax on fossil fuel consumption thereby forcing OPEC to maintain a lower price.

The two major nuclear accidents within a span of seven years (1979 and 1986) once again raised the debate on the safety concern of nuclear energy and its associated social cost. In Europe, strong anti-nuke movement forced many countries to reconsider their ambitious nuclear energy production plans. Moratorium on further expansions was imposed. Moreover, the devastating impact of the Chernobyl disaster dampened the enthusiasm of nuclear lobby that were in desperate search of some moral support. The global warming issue solved their problem and boosted their enthusiasm.

Though coal was found abundantly throughout the world, most of the countries shifted to oil after the 2nd World War. Between 1950s and 1970s, the seven major oil TNCs (the Seven Sisters) who dominated the world global petroleum market till the rise of OPEC power, played the crucial role in this shift from coal to oil. India, with a substantial coal reserve to meet the energy demands for few hundred years had also systematically switched over to petroleum.[4]

Oil TNCs could control world petroleum market over a fairly long period mainly due to their technical edge over others. The technique of coal production was known to almost all over the world but the technique of oil production was unknown to most of the developing countries for a long period. Technology was one of the major tools which helped advanced countries to maintain their dominance over developing nations. With more sophisticated technology, it was easier to control the recipients. TNCs always preferred mega exploration projects, bigger refineries power plants with complex and modern technology.

It was feared that high crude price may compel the developing nations to return back to coal. Indication to this effect was already there. China was planning to fuel its rapid economic expansion almost entirely through its large coal reserves[5]. If other developing countries also followed the Chinese example and relied on coal, the TNCs' control over global energy would be lost substantially. Through sophisticated nuclear technology, developed countries wanted to retain their monopoly on world energy market through TNCs like Asea Brown Bovery (ABB), General Electric (GE), and Siemens AG etc. In contrast, it may be noted that less attention was paid to the development of renewable energy sources like solar and wind, which were technologically less sophisticated, and micro in size. Hence, it was difficult to retain centralized control on these forms of energy sources.

### Europe and Japan: The Active Campaigners

After 2nd World War, Marshal Plan blocked projects for European crude oil production and helped American oil companies to gain control of Europe's refineries. Some $2 bn of total Marshal Plan assistance of $13 bn was for oil imports.[6] In the energy crisis of 1970s, the most affected developed countries were Germany, Italy and Japan. USA with the help of its major oil companies and long term crude supply arrangements with influential OPEC members strengthened its position in the power struggle vis-à-vis Europe and Japan. In the Gulf war (1990-91) also, USA was the ultimate gainer.

To avenge this, few developed countries (mostly from Europe), with practically exhausted oil reserve, tried to put US and its oil corporations into disadvantage. The campaign against global warming offered an opportunity to destabilize the US oil sector. They anticipated that due to global warming campaign, demand for fossil fuel will decline worldwide. The US oil TNCs, which had very large stake in the petroleum industry, would have to bear the cost of warming the globe. At the Rio Earth Summit, Europe took an early lead in advancing plans to help developing nations to combat climate change. The 12 nations European Community pledged to increase aid for environmental projects by $4 bn.[7]

The old conflict between Europe and America to dominate the world economy and politics had certainly made the globe warmer, particularly during and after the Earth Summit at Rio.[8]

## Section II

### The Earth Summit

In 1992, representatives from more than hundred countries assembled at Rio de Janeiro (Brazil) with an objective to save mother earth from further environmental degradation. The 'Earth Summit' at Rio or the World Summit on Sustainable Development (WSSD)-the two names were synonymously used for the gathering – comprising civil societies, policy makers, academics, politicians, corporate representatives, journalists, ministers, presidents, artists and common people, was primarily organized by the UN Commission on Environment and Development (UNCED). In addition to the formation of UN Framework Convention on Climate Change (UNFCCC), several major agreements were enacted at the Earth Summit. These are:

- The Convention on Climate Change – to limit emissions of the greenhouse gases like carbon-di-oxide ($CO_2$) and methane ($CH_4$).
- The Convention of Biological Diversity – to give countries the responsibility for conserving species diversity and using biological resources in a sustainable way.
- The Rio Declaration and the Forest Principles – to set out the principles of sustainable development and pledges to reduce deforestation.
- Agenda 21 – to formulate a plan for achieving sustainable development in the 21st century.

The Rio Earth Summit had put emphasis primarily on three important issues namely: (i) sustainable development; (ii) protection of bio diversity; (iii) global warming. An analysis of the achievements on the above areas during last one decade will indicate the level of seriousness with which such issues were perused in the Earth Summit.

Apart from few ritualistic celebrations on every fifth (Rio+5) or tenth year (Rio+10), the euphoria about sustainable development as created at Rio in 1992, slowly faded out. Burning all their differences, the corporate climate lobbies of Europe and US, cooperated successfully among them to dilute the Rio declarations to their favor by preventing a debate on the need for international regulation on corporations. The absence of such regulation may be considered as the main reason for slowing down the progress in the field of sustainable development policies in the 1990s.

(i) 'Sustainable Development' which is a macro concept with a strong philosophical obligation towards society, demands an all round holistic approach for its proper implementation. The World Business Council for Sustainable Development (WBCSD) was formed in 1995 through a merger of the World Industry Council for Environment (the environmental arm of the International Chamber of Commerce) and Business Council for Sustainable Development. With the formation of WBCSD, the concept of sustainable development was relegated to the back seat paving way to a new concept of 'corporate social responsibility'. It was felt that a 'social compact' among 'shareholders', benefiting from economic activity, and 'stakeholders' suffering from its environmental impact, would be needed. The idea got formalized in the UNCSD held in 1998 which called for an interactive dialogue and voluntary agreements between government and civil society to foster sustainable development.' 'Ecoefficiency' and 'responsible entrepreneurship' were the philosophy behind the voluntary initiatives of the industry.[9] WBCSD with over 140 CEOs of major transnational corporations as its members eventually evolved into the dominant voice on sustainable development and environment. By promoting self-regulation and the concept of sustainable development simultaneously, the WBCSD has helped its members including Dow Chemical Shell, Dupont etc. to green their public image and push for deregulated economic growth and free market globalization.[10]

After ten years of Rio Earth Summit, The Global Environment Outlook 2002, published by UNEP, painted a worsening global trend relating to biodiversity, air pollution, land degradation, chemical emission and wastes, fresh water and regional seas. Analyzing the possible reasons for the widening gap between the efforts being made and the worsening global environmental situation, another UNEP report (2002), published just before the Johannesburg Earth Summit, had identified two main reasons for this.

(a) In most industrial sector, only a small number of companies were actively striving for sustainability;

(b) The improvements were being overtaken by economic growth and increasing demand for goods and services (consumer pressure). It is interesting to note that in the Earth Summit, in a joint statement dated August 28, 2002, The Greenpeace International and WBCSD blamed the governments for 'lack of political will and decisiveness to fulfill their commitments under the Rio agreements including Agenda 21'.[11] Both the parties – the governments and industries got rid of their responsibilities by blaming each other.

(ii) The second important issue of the Earth Summit was the protection of biodiversity. In the absence of adequate safeguards, developing countries, which have protected their biological resources over many centuries, would face unchecked exploitation by the transnational corporations operating in the areas of biotechnology and genetic engineering. The interest of the southern countries with rich biogenetic resources could be best served by a comprehensive global action aimed at conservation and sustainable use of biogenetic resources. In 1992, when the Convention on Biological Diversity (CBD) was formally proposed at the end of the Earth Summit, it was seen as the first decisive step taken by the world community to recognize those rights over their genetic resources were bestowed with the sovereign states. Moreover, it also recognized the need to preserve and maintain the knowledge, innovation, practice of indigenous and local communities. According to CBD, this was to be achieved through fair and equitable sharing of benefits arising from the utilization of such knowledge and practices. CBD was ratified in 1993. Within two years of its ratification, another multilateral agreement was formalized in 1995 under WTO. The Agreement on Trade Related Aspect of Intellectual Property Rights (TRIPS) emphasized on protecting the rights of the inventors. The basic area of conflicts in the two multilateral agreements was as under.

In CBD, the key objective of the treaty was sustainable development where nation states, have sovereign rights over their genetic material. Moreover, the right of the local communities has been recognized for their contribution to the conservation and for any use of genetic material "prior informed consent" of the national states or the local communities who were identified as custodians of the biodiversities was required. Provisions were made in the

CBD for sharing the benefits among the stakeholders for any use of genetic material. It also made it mandatory for the involvement of developing countries supplying the genetic material, in the biotechnological research. TRIPS on the other hand conferred the intellectual property rights over the biotechnological inventors without any consideration as regards the source of the genetic material. As per this agreement, the patent holder need not disclose the source of genetic material on which the patent has been granted and would be the soul beneficiary of the rights arising out of the patent.[12]

Thus, in this important issue of protecting biodiversity, the declaration made at the Earth Summit at Rio has been rendered almost irrelevant with the enactment of a new agreement in TRIPS where the interest of the corporate engaged in biotech and genetic engineering research got the top priority. The granting of patent (May 2003) by the European Patent Office to seed TNC Monsanto on a hybrid Indian wheat variety is a case in point.

Monsanto had patented wheat invented by crossing a traditional Indian variety 'Galahad' with the 'Nam Hal' variety. Monsanto named the 'new' patented variety as 'Galahad 7'. The patent also covered the dough made from the flour and all edible products made by cooking such dough. Gene scientists termed this act as a clear case of theft with the potential to block further breeding of high-quality varieties utilizing this heritage wheat seed. Condemning this, in a strongly worded statement, the international NGO Green peace said, 'Monsanto is targeting and stealing from Indian farmers who have cultured this specific variety of wheat for centuries. This patent demonstrates the urgent need for a general legal ban on the patenting of genes, live organisms and seeds.'[13]

(iii) The other important issue of the Rio Earth Summit, which got maximum priority and media coverage, was the issue of climate change and global warming. The alarming rise of carbon-di-oxide in the atmosphere was identified as the major source of global warming and environmentalists wanted to reduce its emission in the atmosphere by curtailing the use of fossil fuels like petroleum, coal and natural gas. A group of environmentalists sincerely urged to the world leaders to increase the use of renewable energy like solar, wind etc. In the Earth Summit, however, nuclear lobby refrained from highlighting the use of nuclear energy, as an alternative to fossil fuel.

## Kyoto Protocol

UNFCCC, which was created during Earth Summit of 1992 to address the climate change issue, entered into fore in March 1994. It organizes annual meet of the member nations, known as Conference of Parties (COP) In their third annual meet (COP3) at Kyoto, Japan, in 1997, UNFCCC agreed to a Protocol on reduction of greenhouse gas emission.

The Kyoto Protocol was negotiated to transform the UNFCCC goals into legally binding policies. The protocol set individualized carbon-di-oxide emissions targets for each of the Annex I parties (industrialized countries including 'economies in transition' like Russian Federation), which would add up to a total reduction of 5.2 percent below the level of their collective emissions in 1990 during the commitment period, 2008-2012. The developing countries have been grouped as Non-Annex I countries. At present, Non-Annex I countries have no emission reduction target. The US target is 7 percent below 1990 levels; the European Union target is 8 percent. It came into force on 16th February, 2005, 90 days after Russia's ratification. The protocol required ratification by 55 parties including enough Annex I parties to account for at least 55 percent of the world's carbon-di-oxide emissions in 1990. With Russia's ratification, 61 percent of all industrialized countries' 1990 emissions will have been accounted for. As of 6th February, 2005, 141 countries ratified the Protocol with the prominent exception of USA and Australia who emit substantial greenhouse gases in the atmosphere. Upon entering into force, US companies with factories in countries party to the Kyoto Protocol will have to reduce emissions to comply with established target levels of the concerned country.

To meet the required 55 percent threshold, either the United States or the Russian Federation had to ratify the treaty. In 2004, the Russian Federation announced that it would ratify the Protocol in exchange for EU support for Russia's admission to the WTO. Commenting on this, the *Economic and Political Weekly of India* in their editorial, wrote "... the manner in which the EU, for instance, has gone about 'persuading' Russia to sign the protocol give rise to suspicions that this is now less about environment and more about business"[14].

The Kyoto Protocol established several innovative mechanisms to reduce the global costs of compliance. These include an emissions-trading scheme, allowing countries to buy and sell the right to emit certain quantities of greenhouse gases when reductions in other countries can be achieved more cheaply, and a "clean development mechanism" (CDM) that rewards Annex I parties for generating investments in less-polluting energy systems in developing countries.

### US Position

President Clinton signed the Kyoto Protocol in 1998, but President Bush reversed the earlier US position and effectively ended US involvement in 2001 because he thought that in its current form, the Kyoto Protocol would "severely damage the United States' economy". The Bush administration also had reservations to ratify a treaty that did not equally obligate developing countries including India and China.

In 1992, during the Earth Summit at Rio, senior Bush (the father of present president Gorge Bush) of Republican Party with strong linkage with Texas oil lobby was the president. Naturally, he was a 'reluctant participant' to the Earth Summit. After the defeat of Sr Bush, Bill Clinton of the Democratic Party became the President. After coming to power, President Clinton proposed an energy tax in 1993 by taxing oil use at a higher rate than other fuels. Renewable sources of energy would have been tax-free. Energy industry lobbies worked against it intensively and the proposal failed to pass Congress.[15]

Again in 1997 when USA signed Kyoto Protocol, President Clinton was in power but he did not get enough time to ratify the same within his presidential tenure. For three years, between 1997 and 2000, when Clinton wanted to ratify the treaty, the oil, coal, utility, and automobile industries had generated studies predicting economic disaster for the United States if the Protocol were implemented. Control of both the House and Senate by Republicans closely allied with the oil, coal, utility, and automobile industries made domestic action to reduce carbon-di-oxide emissions difficult. This difficulty was compounded by alliances between a significant number of congressional Democrats and mining and autoworkers' unions. After Clinton's departure, when the present Republican

President George Bush came to power, the oil lobby returned back to Whitehouse with full strength. Clinton's decision to be a party to the Kyoto Protocol was revised and in March 2001, US President Bush decided to withdraw from further talks on Kyoto Protocol.[16]

After four years of US withdrawal, President Bush maintained the same attitude towards Kyoto Protocol. In July 2005, prior to G-8 Meeting on global warming, Mr. Bush commented "The Kyoto treaty would have wrecked our economy, if I can be blunt".[17] This statement again corroborates our suspicion that from its very inception, the concept of 'global warming' was pregnant with political overtonnes.

## $CO_2$ and Global Warming – Scientific Findings, Few Questions

In this sub-section, first we shall compile few news items on global warming to assess the recent developments in this field of knowledge. Then we shall summarize two articles by Riggs (1998) and Appenzeller (2004) which have raised few pertinent questions on this debate. Let us begin with few recent news items as under.

- 'Nearly all climate scientists today believe that much of Earth's current warming has been caused by increase in the amount of green house gases in the atmosphere, mostly from burning of fuel', commented Mr. Ralph Cicerone – the President of the National Academy of Science, USA, while testifying before a Senate Commerce sub-committee on Global Climate Change.[18]
- Intergovernmental Panel on Climate Change (IPCC) in its 3rd report had furnished an estimate on the safe global GHG concentration level at around 550 ppmv by 2100. In this regard, the Special Committee on Future Framework for Addressing Climate Change Constituted by the Ministry of Energy, Trade and Industry, Japan had observed that 'quantitative analysis of the impact on the climatic system that will occur as a result of specific temperature rises is insufficient, and most analyses have been qualitative thus far. Given the state of current scientific knowledge, attempt to engage in negotiations towards gaining international agreements on specific values would not be constructive.'[19]

- Global warming could significantly increase rainfall in Saharan Africa within a few decades, potentially ending the severe droughts that have devastated the region, a new study suggests. The discover was made by climate experts at the Royal Meteorological Institute in De Bilt, the Netherlands, who used a computer model to predict changes in the Sahel region – a wide belt stretching from the Atlantic to the horn of Africa that includes Ethiopia, Somalia and Djibouti. Some scientists suspected that global warming might increase rainfall in the region, causing the so-called greening of the Sahara, but these are the biggest predicted increases so far. Scientists say the increased rainfall could "strongly reduce the probability of prolonged droughts."[20]
- Gerry Stanhill, an English scientist working in Israel, comparing four decades of Israeli sunlight records from the 1950s, observed a 22% drop in the sunlight reaching Earth's surface in that period. Then, he collected similar records from other countries too. Though the drop in sunlight varied from place to place, overall the decline amounted to 1-2% globally per decade between the 1950s and the 1990s. In his research paper published in 2001, he called the phenomenon as "global dimming". Scientific community did not pay much attention to this initially. Later, independent study by other scientists also claimed similar phenomenon. Now, it has been revealed that 'dimming' is primarily caused by air pollution. Burning of coal, oil and wood produce carbon-di-oxide and other pollutants like tiny particles of ash, soot, sulphur compounds etc. While carbon-di-oxide has been identified as one of the greenhouse gases that cause global warming, the other pollutants like tiny particles etc. are responsible for global dimming. By default, these two different types of pollutants are affecting our climate in a completely opposite way, thus balancing their individual impacts. This new finding demands a thorough re-examination of the climate change model which predicted a steep rise in global temperature due to greenhouse effects.[21]
- Between now and 2050, deep underground storage of carbon-di-oxide could prevent between 20 and 40 percent of global emissions, largely responsible for global warming, according to a report by UN scientists obtained by AFP. Between 220 billion tonnes and 2,200 billion tonnes

of $CO_2$ could be economically stored underground in geological structures such as empty oil and gas fields and in the deep ocean between now and 2100, the report said.[22]

If we refer to the first news item as above, the beginning two words of the statement ('nearly all') indicate, *till now*, a section of climate scientists *are* skeptical about the global warming theses, the way it is being propagated. In the second news item also we find, a section of Japanese Scientists had expressed similar reservation. According to them, the current scientific knowledge in this field was insufficient.[23] The next two news item tells about some positive aspects of 'global warming' especially to balance 'global dimming' and the last one indicates about a possible solution for excess $CO_2$ in the atmosphere.

Riggs (1998),[24] in his article titled 'Global Warming: Divided Science and Unfounded Policy' raised few important questions on which the 'global warming' thesis has been developed. Below, we summarize the major findings.

- The greenhouse effect is a naturally occurring phenomenon: the sun's energy, in the form of solar radiation, enters the earth's atmosphere and is converted into heat. Greenhouse gases – water vapor, carbon-di-oxide, and methane – absorb this heat and further warm the earth's atmosphere. The general theory of the greenhouse effect says: holding everything else constant, an increase in greenhouse gas emissions increases temperature. Global climate, like weight loss, is determined by many factors. It is the interplay of these other factors, known as "feedback effect" that fuels the global climate change debate. Changes in water vapor and cloud cover are two of these feedback effects.
- We know so little about cloud feedback and water vapor feedback, which is probably the most important feedbacks to carbon-di-oxide, the safest thing is to assume that the earth will continue to do what it does best: that is, reject excess heat.
- A report released in the summer of 1996 by the United Nations' Intergovernmental Panel on Climate Change (IPCC) has been the basis for what subsequently has been called a "virtual consensus among scientists" on global warming. Immediately after release of the IPCC report, Frederick Seitz, President emeritus of Rockefeller University in

New York City and Chairman of the George C. Marshall Institute, wrote, "In my more than 60 years as a member of the American scientific community, including service as president of both the National Academy of Sciences and the American Physical Society, I have never witnessed a more disturbing corruption of the peer-review process than the events that led to this IPCC report."

- In 1990, the IPCC (established in 1988) predicted a 0.3 degree centigrade increase in earth's temperature per decade. Just two years later, the IPCC lowered this estimate slightly after learning more about feedback effects. The IPCC's 1996 estimates showed an increase of 0.18 degree centigrade per decade. One might infer from these results that a downward trend in predicted temperature estimates is evident as scientists learn more about the climate system and incorporate this knowledge into their models.
- In 1995, some 100 independent scientists expressed their skepticism about global warming in the Leipzig Declaration on Global Climate Change, written at an international symposium in Germany. The Leipzig Declaration summarizes their doubts thus: "It has become increasingly clear that contrary to conventional wisdom, there does not exist today a general scientific consensus about the importance of greenhouse warming from rising levels of carbon-di-oxide. On the contrary, most scientists now accept the fact that actual observations from earth satellites show no climate warming whatsoever. And to match this fact, the mathematical climate models are becoming more realistic and are forecasting temperature increases that are only 30 percent of what was considered the "best" value just four years ago." The Leipzig Declaration urged caution in moving forward with greenhouse gas emission-reduction policies based on limited current knowledge.
- In a study (January 1998) on 'Environmental Effects of Increased Atmospheric carbon-di-oxide,' the authors (Arthur S Robinson *et al*) did not find any evidence to support the theory that increased carbon-di-oxide were likely to cause catastrophic changes in global temperatures or weather. In fact, they analyzed the real possibility that more carbon-di-oxide in our atmosphere would accelerate plant growth rates, and, since animals depend on plants, animal life might also flourish. They concluded: 'We are living in an

> increasingly lush environment of plants and animals as a result of the carbon-di-oxide increase. Our children will enjoy an earth with far more plant and animal life than that with which we now are blessed.'

We may recall, in early 1990s, Dr Isdo also came to a conclusion (please refer to Section I) strikingly similar to the above conclusion of Prof Robinson and his colleagues.

Appenzeller (2004)[25] filed a report titled 'The case of the missing carbon' in the online edition of National Geographic Magazine. For this article, the author was awarded with the 2005 Walter Sullivan Award for Excellence in Science Journalism. As the name of the article suggests, here he tried to address few basic questions mainly on the linkage between carbon-di-oxide and rise in global temperature. Here is a brief summery.

- Humanity dumps roughly 8 billion metric tons of carbon, each year, into the atmosphere. Of this, 6.5 billion tons are from fossil fuels and 1.5 billion from deforestation. But less than half that total, 3.2 billion tons, remains in the atmosphere to warm the planet. Where is the missing carbon?
- Forests, grasslands, and the waters of the oceans must be acting as carbon sinks. They steal back roughly half of the carbon-di-oxide we emit and delaying the effects on climate. Each acre of the forest has been taking roughly three-quarters of a metric ton of carbon out of the atmosphere annually, doing its humble part to counteract greenhouse warming. Plants need carbon-di-oxide to grow, and scientists have found that in laboratory chambers, well nourished plants bathed in high carbon-di-oxide air show a surge of growth. Thus, out in the real world, it seemed, plants would grow at a faster rate as carbon-di-oxide built up in the atmosphere, stashing more carbon in their stems, trunks, and roots and helping to slow the atmospheric buildup. Such a growth boost could, for example, turn mature tropical forests—which normally do not soak up any more carbon than they give off—into carbon-di-oxide sponges.
- The most fitting end for the carbon that human beings have tapped from the Earth—in coal, oil, and gas—would be to send it back where it came from into coal seams, old oil and gas fields, or deep, porous rock formations. Not only would that keep the carbon out of the atmosphere, but the

high-pressure injection could also be used to chase the last drops of oil or gas out of a depleted field. Geologic sequestration, as it is called, is already under way. One field in the North Sea, for example, yields gas that is heavily contaminated with natural carbon-di-oxide. So before shipping the gas, the Norwegian oil company Statoil filters out the carbon-di-oxide and injects it into a sandstone formation half a mile below the seafloor. The North Sea project seems to be working well eight years after it began. Seismic images that offer views beneath the ocean floor show that the thick layer of clay capping the sandstone is effectively sealing in the six million tons of carbon-di-oxide injected so far. That is encouraging news for researchers who are working on schemes that would allow humanity to keep burning fossil fuels without dire consequences for climate.

- Researchers at Princeton, for example, are exploring a technology that would take the carbon out of coal. In a multistep process, coal would react with oxygen and steam to make pure hydrogen, plus a stream of waste gases. The hydrogen could be burned to produce electricity or distributed to gas stations where hydrogen-powered cars—emitting nothing but water vapor—could fuel up. The waste, mostly carbon-di-oxide but also contaminants that coal-burning plants now emit, such as sulphur and mercury, would be buried. The scheme, says Princeton energy analyst Robert Williams, "could make coal as clean as renewable energy, and you can exploit the low cost of coal." Or may be the future lies in fields of solar panels, armies of giant wind turbines, or a new generation of safe nuclear reactors.

The analysis of the above news items and two articles suggest once again that 'global warming' issue has remained as in the past an issue of divided opinion where minority scientific arguments are drowned with a cacophony of majority views. Is it because the climate change issue has a huge commercial stake attached to it?

## The Strategic Importance of Weather – Its Changing Role

The study of climate was integral part of every civilization for developing better survival strategy for the mankind. In the history of all civilizations, we find, climate was considered as 'given'. Those who could utilize it to their advantages could grow faster. No civilization, till the middle of last century, did modify and control the climate for their strategic advantage. Climate studies entered into the strategic discourse of military academy in the early sixties (1963), when a breakthrough in 'dynamic seeding' of cloud at Florida by the US Naval Weapons Centre was reported. During the cold war period, defense scientists and strategists were engaged in designing 'weather weapons' of different types and magnitudes. Projects were undertaken for developing techniques of controlling and directing natural phenomena like earth quakes, tsunamis, cyclones, desertification, flash floods, tidal waves, lighting etc.

Seshagiri (1977)[26], in his famous book 'The Weather Weapon' wrote: 'From a dubious beginning in 1839, the science of weather modification and control (WMC) passed through stages of theoretical formulation of rain formation, the spectacular discoveries of Langmuir and Schaefer, the control of hurricanes and finally the reported use of the technology in weather war over indo-China.' Further, the author claimed that 'In the background of the considerable interest shown by the military in some countries, and the fact that many well funded projects are under the cloak of secrecy, one may confidently assert that WMC is now a full-fledged technology.'

By early seventies, the threat of weather weapon became very loud and clear. It offered a cheaper option even before a poor country like India to use it as a weapon of mass-devastation. To control the propagation of such a deadly weapon, UNEP at its Governing council in Nairobi in March 1974, adopted a resolution on the code of conduct on the man-induced weather modification. And in 1976, the International Pugwash Movement appealed to ban weather weapon. But experiments in this new area of study continued in different countries.

In the current decade, the melting of arctic ice due to global warming is a serious concern to the leaders and environmentalists across the globe. In the early seventies, due to its high potential to adversely affect the economies and livelihood of island countries and coastal regions, 'melting of arctic ice' was considered as a potential weather weapon by the Soviet Union and the US. Seshagiri in his book has mentioned about two such projects – namely Project PECHORA (USSR) and Project BEARING – DAM (USA). The first one was an ambitious project for reversing the flow of river Pechora so as to warm the Arctic by reducing fresh water inflow and thereby setup a chain reaction to warm the Earth. The second project contemplated the construction of a dam blocking the Bering Strait between Siberia and Alaska and pump out water from the Arctic Ocean into the North Pacific so as to draw warmer water from the North Atlantic into the North Arctic. The project was on hold as the scientists were not sure whether the process would melt the arctic ice or trigger another ice age.

In this new century, the national economies are getting increasingly integrated to the global economy. TNCs now play more important role than the sovereign nation states in developing rules and regulations of different multilateral treaties pertaining to important issues like climate change, intellectual property rights etc.[27]. The powers of the nation states are eroding very fast. TNCs with the help of mainstream global media, on which they have almost total control, could systematically establish or suppress any 'scientific' view, which are important for their strategic planning. Under this changed power structure, old strategies and tactics are bound to change. 'Weather' which had a strategic importance as 'weapon' to a nation state in cold war period, is bound to have a strategic importance of different type (as 'economic good') to a TNC dominated world order of this century. Now the emphasis is more on economics than politics. The paradigm shift is very clear and distinct.

## Role of Global Corporations in the 'Climate Change' Business

To understand the role of global corporations in the climate change debate, we have heavily relied on different publications of Corporate Europe Observatory (CEO). Below, we narrate few of their important findings.[28]

- Despite the strategic differences between corporate climate lobbies on both sides of the Atlantic, they work closely together in a number of powerful international coalitions. The US lobby groups campaigning against ratification of the Kyoto Protocol work hand in hand with Europe-based groups that pursue a more 'constructive' image to shape the Kyoto 'rulebook' in the interests of transnational corporations. Two groups operating at the international level, the World Business Council for Sustainable Development (WBCSD) and the International Chamber of Commerce (ICC), have been very diligent in lobbying for the right climate for industry, through their mastery of the doctrine of corporate environmentalism. The main success of this strategy has been to portray corporations as part of the solution, and not the problem. At the UN climate negotiations, these groupings coordinate their lobbying efforts to achieve optimal political impact. The US lobby groups campaigning against ratification of the Kyoto Protocol work hand in hand with Europe-based groups that pursue a more 'constructive' image to shape the Kyoto 'rulebook' in the interests of transnational corporations.

- Both the WBCSD and the ICC tout the Clean Development Mechanism (CDM) as the most promising solution to the climate crisis. It will provide ample opportunities for TNCs to win new markets in Southern countries as the carbon economy continues to take hold.
- The two groups have nearly identical demands for the climate debate. They want to "bring developing countries on board," promote voluntary industry agreements as the preferred regulatory option, oppose any restrictions on the use of the flexible mechanisms, and include nuclear, coal, large hydroelectric and plantation forestry as carbon sinks under the CDM and Joint Implementation (JI). The ICC is also particularly keen on self-monitoring by industry and on avoiding binding non-compliance measures such as fines or penalties.
- Both groups have kept busy on various fronts since Kyoto. The WBCSD runs the 'International Business Action Plan on Climate Change' which promotes CDM projects among its corporate members and Southern countries, working together with the UN Development Program (UNDP) to facilitate the involvement of the private sector in these projects. It is simultaneously involved in three major attempts to accelerate the creation of global $CO_2$ emissions trading markets: one coordinated by the UN Conference on Trade and Development (UNCTAD), the World Bank carbon trading fund and the newly established International Emission Trading Association (IETA). IETA promotes corporate emissions trading and counts major polluters such as BP Amoco, Royal Dutch Shell, Statoil, Endesa and Tokyo Electric Power among its members.
- The ICC's enormous resources give it a formidable lobbying capacity. Its Working Group on Climate Change consists of over 70 executives from

different corporations. The ICC plays a coordinating role for business lobby groups at UN negotiations, organizing daily information exchanges and strategy meetings as well as meetings with UN officials. Indicative of the ICC's privileged political access was an international meeting in Dakar, Senegal just two weeks before the Buenos Aires Climate Summit. A 30-person delegation from the ICC and WBCSD joined representatives from Shell, LaFarge, Texaco, Mobil and Chevron in discussions with energy and environmental ministers from more than 20 African countries. The agenda was to tempt these Southern governments with promises of technology transfer and foreign investment in exchange for political support for the CDM. A follow-up meeting was organised in the spring of 1999.

- During COP-4 in Buenos Aires, the ICC, together with Keidanren, the WBCSD, the USCIB, the Worldwide Fund for Nature (WWF) and the United Nations Environment Programme (UNEP) organized a 'Workshop on Voluntary Initiatives on Climate Change'. The ICC's lobbying contingent of over 100 business representatives took the lead in the heavy industry lobby that contributed to postponing decisions on the implementation of the flexible mechanisms until COP-6. Both the ICC and the WBCSD have made good use of this interlude to continue lobbying industrialized and developing countries to shape the Kyoto rulebook in the most business-friendly manner possible.
- Another vehicle for corporate power within the climate debate is the Transatlantic Business Dialogue (TABD). Through the TABD, EU and US-based corporations develop policy recommendations which governments on both sides on the ocean have committed themselves to implementing. While mainly a tool for deregulation and the creation of a transatlantic free trade area, the TABD has, since 1997, had an active 'issue group' on climate change. This working group is made up of representatives of European and US corporations, with a co-chair from each side of the Atlantic.
- The greenhouse gas brokers have organized themselves in lobby groups such as the Emissions Marketing Association (EMA). Working with the slogan of "Serving the International Emissions Trading Community," EMA brings together (in 2000) over 140 consultancies and corporations from around

the world, including Mitsubishi, Cargill; Enron and Dow, as well as lobby groups like the Global Climate Coalition. The EMA, which publishes the monthly newsletter, 'The Emissions Trader', lobbies against any kind of restrictions or limitations on the use of the emissions trading mechanisms.

On March 2, 2006, India and USA have entered into a civilian nuclear deal where India promised to separate its civilian and military installations in return for uninterrupted supply of uranium and access to advanced nuclear technology to fuel growing energy needs.[29] However, to make it operational, the US Congress must ratify it. It is reported that sensing the potential of the deal, (The Indo-US deal has far reaching consequences on the composition of future fuel mix of India's energy supply. It is feared that India, in future, would rely more on imported uranium than on abundantly available indigenous coal for power generation. With the signing of the deal, $100 billion business opportunities have been created for US firms in energy sector alone. (The Economic Times, 3.3.2006) a strong contingent of the US business community under the leadership of US chamber of Commerce, has already launched the second phase of its strategic initiative to ensure that the 'deal' gets through in the US congress. The chamber has planned to host 'coalition for partnership with India', to marshal a broad public advocacy campaign. At least three different lobby firms are working simultaneously to seek the nod of US congress on the nuclear deal.[30]

The above findings clearly indicate that the 'sovereign power' of the 'state' in regulating and managing vital issues like environment has been eroded through the conclusion of multilateral treaties and protocols where these corporations played important role in formulating rules and regulations for their own vested interest. In such a commercial model of 'development', the basic goal of 'sustainable development' has no place to play. Systematically, it has been pushed to oblivion.

## Section III

### Net Effect on Energy Consumption

In the previous section, we discussed about the third and core issue of the 1992 Earth Summit, the issue of global warming and the objective of restricting carbon-di-oxide emission in the atmosphere through reduced consumption of fossil fuel and increased usage of renewable energy. Global leaders (barring few exceptions) showed enthusiasm about the move. Energy consumption data unfortunately do not reflect any such serious effort by world leaders including leaders from Europe, who lead the bandwagon. The tables in the annexure will bear the truth. Between 1990 and 2003, the world petroleum consumption grew by over 20%. In case of Western European countries, the growth rate was over 12% and during the same period, growth rate in Japan was nearly 7%. However, the growth rate in Eastern Europe and former Soviet Russia, was over – 44%. But the negative growth rate was primarily due to slowing down of the economy. In case of coal, between 1990 and 2001, the total world consumption remained stagnant. While consumption in Europe declined (switching to natural gas), consumption in developing countries increased. And consumption of industrialized countries taken together also remained stagnant.

In 2002, the EU (25 member) consumed 18% of the world's total energy. In the same year, USA, China, Japan and Russia consumed 24%, 10%, 5% and 7% respectively. Others consumed rest 36%. In that year, EU energy basket consisted of 40% oil, 22% natural gas, 16% coal, 13% nuclear, 4% hydro and only 1% renewable other than hydroelectric power. In the post Rio decade, natural gas was the fastest growing fuel in EU mainly at the expense of coal whose share declined from 20% in 1991 to 16%. Apart from environment concern, the easy availability of natural gas supply from Algeria, Norway, and Russia by pipeline and LNG import from Nigeria was responsible for this switch from coal to gas. In 2002, Poland produced 178 million short tons of coal. It is estimated that due to increased Polish production, the share of coal in EU energy basket would increase to 30% in the near future. The growth of oil and coal consumption in two other major energy-consuming countries of the world, namely, the USA and China also indicate the hollowness of the Rio declaration. Between 1990 and 2001, coal consumption (million short tons) in USA increased from 903 to 1060, an increase of over 17%. The corresponding figures in China were 1,124 and 1,383. And between 1990 and 2003, the petroleum consumption (ref: Annexure I) in those two countries increased by 17.92% and 141.68% respectively. Future projections by OPEC (2004) also do not portray any encouraging scenario up to 2025.[31]

| Table 1: Share of Total Primary Energy Sources, 2003 | | | | | |
|---|---|---|---|---|---|
| **Country** | **Oil** | **Gas** | **Coal** | **Nuclear** | **Hydro, Renewable and Wastes** |
| USA | 40.4 | 22.8 | 23.3 | 9.0 | 4.5 |
| EU-25* | 40.2 | 22.4 | 15.6 | 13.0 | 5.2 |
| China | 19.1 | 2.5 | 60.3 | 0.8 | 17.3 |
| Russia | 20.5 | 53.4 | 16.8 | 6.2 | 3.1 |
| Japan | 49.7 | 13.7 | 20.8 | 12.1 | 3.7 |
| Brazil | 44.4 | 6.7 | 6.9 | 1.8 | 40.2 |
| India | 22.4 | 4.2 | 33.2 | 0.8 | 39.4 |

# For 2002. The total is slightly short of 100% may be due to various adjustments to be made while compiling the aggregate data

*Source: http://www.iea.org/Texbase/pamsbd/JRECPIC/Shares%20of%20TPES%202003%20-%2..; Country Analysis Briefs – Regional Indicators: European Union (2004), www.eia.doe.gov*

Kyoto Protocol lost its relevance the day when USA, the world's largest green house gas emitter decided to pull out of it. Analysts have calculated that, at this present form, the actual reduction of carbon-di-oxide under market driven mechanism will lead to a reduction of 0.1% instead of a miserable target of 5.2% of 1990 level.[32]

Voicing its concern on slow initiatives towards addressing the climate change issue, the environmental advocacy group—The Friends of the Earth—alleged in a recent report (October 2005) that the World Bank had rarely considered climate change strategies when lending for projects in developing countries. According to them, the Bank also failed to adequately fund and create policies to push development of clean energy.[33]

Unable to tame the major carbon polluters of the developed North, the environmental groups are now targeting the Southern states –the softer options.

## Who Gains?

In this entire 'climate change' debate, there are two groups of immediate gainers (a) the corporate groups engaged in environment business (b) the nuclear energy lobby.

(a) **The Corporate Group:** The climate related issues have led to two major global treaties namely the Montreal Protocol (1987) and the Kyoto Protocol. The former, as we know, vowed to eliminate ozone depleting substances (ODS) from the earth. Accordingly, CFC (Chloro Floro Carbon) – a man made 'miracle' gas was phased out from our refrigerators and air conditioners. A very conservative estimate indicates that Indian domestic consumers had to pay $1.2 billion to the refrigeration companies for converting domestic refrigerators to a non ODS one.[34] In a global scale, a huge non-ODS market was thus created where non-ODS products/ technology worth billions of dollar was transacted.

Like its predecessor Montreal Protocol, Kyoto Protocol has also created a new market through 'emission trading' (trading 'emission allowances' among developed nations), 'joint implementation' (transferring 'emission allowances' between developed nations, linked to specific emission reduction projects) and the 'clean development mechanism' (CDM). Through CDM, developed nations are allowed to achieve part of their reduction obligations through projects in developing countries that reduce emission or 'fix' or sequester $CO_2$ from the atmosphere.

It is claimed that the developing countries which have ratified the Kyoto Protocol but do not have any emission reduction obligation, will be benefited by CDM in two ways[35]:

i) It will help developing countries to achieve sustainable development.

ii) By allowing developed countries to take credits from emission reducing projects undertaken in developing countries, the developed countries will be provided with the flexibility of achieving the emission reduction targets. At the same time, the developing countries would be financially benefited by selling the certified carbon reduction units (CERS) to the purchasers from the developed countries.

However, it is difficult to estimate how much benefit the developing countries would get out of CDM due to its various restrictive clauses.[36] For example, decision was taken in COP6 (2000) that afforestation and reforestation were the only eligible land use activity under CDM for the first commitment period (2008-2012). Moreover, only areas that were not 'forest' on 31st December, 1989 were likely to meet the CDM definition of afforestation and deforestation.[37] Again, to become eligible for payment under CDM, the claimant must follow a very strict procedure. And the actual payment against carbon trading, if any, would be made only in 2007 after the 'international transaction log' (ITL) is put in place. Corporate lobby were also successful in rejecting the proposals for public involvement in the CDM. This is why the countries like Papua, New Guinea and Costa Rica who saves the famous 'rain forest' which act as a natural carbon sink are not eligible for any monetary assistance in the CDM.[38]

The CDM was intended to facilitate the transfer of technology for energy efficiency measures to Southern countries. The umbrella group of countries consisting of the US, Japan, Australia, Canada, Russia and Norway wanted all technologies, including nuclear energy and the whole range of carbon sinks, to be included in both CDM and JI. Monsanto – the US based transnational corporation, claimed that massive amount of CO2 would remains stored in the soil if farmers reduce or stop ploughing and instead, use its herbicide and genetically modified crops. Monsanto also expects to earn huge profit with genetically modified plants and trees that take up or store carbon more effectively.[39] In near future, to earn emission credit under CDM project, genetically engineered (GE) forest may come up in different parts of Southern countries evicting local people from their ancestral land.

A new industry has emerged out of Kyoto Protocol by selling and buying of 'hot air'. In 1999, the emission trading market was $50 billion which is likely to reach $13 trillion by 2050. Through 'emission trading' Russia, Ukraine and Kazakhstan, the $CO_2$ emission of which are some 45% below 1990 level, would be the major immediate beneficiaries. These countries may earn $30 billion per year through trade in 'hot air'. The World Bank had developed a prototype Carbon Fund even before finalization of Kyoto Rule Book to mobilize funds from Northern governments and corporations and pumps them into projects in Southern countries. The World Bank in return pays out carbon credits, related to the emissions reductions achieved, as dividend to the investors. Among the first investors in the Bank's Carbon Fund were the Dutch government and corporations such as Mitsubishi and Shell.[40]

In addition to a new market for trading in hot air, the Kyoto Protocol in general and CDM in particular, has opened up the huge potential for sale of 'clean technology' to the vast Southern countries eager to participate in the CDM projects. With the formation of Carbon Fund by the World Bank where mega corporations have started to 'pump' in money for Southern countries and the constant monitoring by the global civil societies like the 'Friends of the Earth' on the project financing patterns of the World Bank, there would not be any dearth of fund to finance the CDM projects of the developing countries.

The terms 'carbon emission', 'carbon trading' etc. have become synonymous to CDM projects. But a recent report narrates a different picture. The report[41] claims, destruction of HFC-23, a by product generated by HCFC-22 (a gas used in refrigeration) is globally the most popular CDM scheme. The UNFCCC secretariat estimates that if 30-odd plants in developing countries already manufacturing HCFC-22 – apply for CDM benefits by incinerating HFC-23, more than 100 million CERS per year could be generated by 2012 (the time by which developed countries must meet their first set of commitments to reduce global warming) This is a large slice in the CDM market, which is 50-180 million tonnes of carbon equivalent, or 50-180 million CERS, per year. In terms of the projects registered and under UNFCCC validation, HFC-23 accounts for almost 12 million CERS a year, roughly 24 percent of all CERS being sold.[42] This explains why some transnational corporations dealing in refrigeration products had taken double standards during implementation process of Montréal Protocol. In developed countries, the transnational corporations had switched to more benign hydrocarbon (HC) technology but in developing countries like India, they preferred HCFC or HFC knowing fully well that mere destruction of HFC-23 would earn huge commercial benefits under a CDM project in future. It may be noted that in most of the developing countries, when the ODS phase out as per Montreal Protocol was notified Kyoto Protocol (1997) had been already signed and by 2001 at COP 7, the CDM procedures were also finalized. For example, in India, as per the ODS Rule 2000, the phase out process was to start from January 1, 2003.

Analyzing two projects undertaken in India under CDM, the above report commented that sustainable development was not an important goal to such projects. The CDM Auditors attached high weightage on proper 'documentation of benefits' rather than the actual benefit of the environment through such project. Now, the commercial potential of 'global warming' indicates that the global environment issue has become an industry hobby-horse made for the industry, by the industry.

However, it is heartening to note that few Northern civil societies have taken initiative to correct these aberrations. For example, The Helio-International has extended its methodology and indicators to provide a tool for countries to select good development projects genuinely helping the South. This facilitated the creation of South-South-North (SSN)[43] and the matrix they developed during the pilot project was SSN1. SSN has also contributed to the development of the International Gold Standard label which ensures the highest standards of practice throughout CDM project development and implementation. Host countries and their designated national authority (DNA) may use the SSN matrix to ensure that they get good projects.

(b) **The Nuclear Lobby:** The other beneficiaries of the global warming issue is the nuclear lobby which as anticipated earlier in Section I, has consolidated its position in the post 'Earth Summit' period. After signing of the Kyoto Protocol (1997), the nuclear lobby representing European Atomic Forum (FORATOM), the European Nuclear Society (ENS), and their associate organizations in US, Canada and Japan worked hard to include nuclear energy in the Kyoto rulebook as a solution to the de-carbonization of the energy supply. Inclusion of nuclear energy in the CDM would have served two important purposes. (i) New commercial opportunities would arise due to its potential to earn emission credits (ii) inclusion in the Kyoto Protocol would have increased the credibility and acceptability of the nuclear technology to energy policy makers. Though the strong nuclear lobby has failed to include (till COP11) nuclear energy in Kyoto Protocol, the aggressive propaganda has paid dividends. By 2001, the nuclear power accounted for 35% of the total electricity in the 15 member EU.[44] The table below will indicate the growth of nuclear energy in different countries between 1990 and 2004.

**Table 2: Nuclear Consumption (Million Tonnes Oil Equipment)**

| Region/Country | 1990 | | 2004 | | Growth (%) in 2004 over 1990 Consumption |
|---|---|---|---|---|---|
| | Consumption | Share on on Total(%) | Consumption | Share on Total(%) | |
| USA | 137.4 | 30.32 | 187.9 | 30.10 | 36.75 |
| France | 71.1 | 15.69 | 101.4 | 16.24 | 42.62 |
| Germany | 34.5 | 7.61 | 37.8 | 6.05 | 9.57 |
| Spain | 12.3 | 2.71 | 14.3 | 2.29 | 16.26 |
| Sweden | 15.4 | 3.40 | 17.3 | 2.77 | 12.34 |
| United Kingdom | 14.9 | 3.29 | 18.1 | 2.90 | 21.48 |
| Italy | – | – | – | – | – |
| Norway | – | – | – | – | – |
| Poland | – | – | – | – | – |
| Russian Federation | 26.8 | 5.91 | 32.4 | 5.19 | 20.90 |
| European Union | 175.5 | 38.72 | 223.4 | 35.78 | 27.29 |
| Brazil | 0.5 | – | 2.6 | 0.4 | – |
| India | 1.4 | 0.31 | 3.8 | 0.61 | 171.43 |
| China | – | – | 11.3 | 1.81 | – |
| Japan | 44.3 | 9.77 | 64.8 | 10.38 | 46.28 |
| Total Asia Pacific | 65.2 | 14.39 | 118.9 | 19.05 | 82.36 |
| **Total World** | **453.2** | **100.00** | **624.3** | **100.00** | 37.75 |

*Source: BP Statistical Review of World Energy 2005.*

Table 2 not only indicates about the growth of nuclear energy, it also highlights the fact that, unlike in previous decades when nuclear energy consumption was primarily confined to North America, EU and Japan, during 1990 and 2004, developing countries like India, Brazil and China have made their presence felt in the growing nuclear energy club. In absolute terms, though India's contribution was not very significant but in terms of rate of growth, in that period, it grew by 171.43% – the highest growth rate among all the countries. It is likely to increase further due to the renewed interest taken by the present UPA government to boost nuclear energy.[45] The Indo-US deal signed in March 2006 is a case in point.

After the ratification of the Kyoto Protocol in February 2005, except in Germany, there were signs of 'mini-boom' on nuclear energy in Europe, commented a Bloomberg.[46] Some recent developments would indicate that the 'boom' is not confined to Europe only.

- In Finland, the first new nuclear station in Europe since 1986 disaster in Chernobyl has been ordered.
- In France, Electricite de-France SA declared that it would build the first of a new generation of nuclear-power stations. The shares of Areva SA, the world's largest maker of such plants, have surged to 330 euros (early June, 2005) from 127 euros in February 2003.
- In June 2005, China National Nuclear Corporation disclosed its plan to invest $48 billion in next 15 years in building nuclear-power plants.
- In US, NuStart LLC, aims to license the first US nuclear power plant in 30 years and are looking for potential sites.
- In November 2005, Tony Blair, the Prime Minister of Britain, unveiled before the Confederation of British Industry conference in London, a wide ranging review of Britain's energy needs which is expected to pave the way for a new generation of nuclear power stations.[47]
- ElBaradei – the Chief of International Atomic Energy Agency (IAEA) and IAEA have been jointly awarded with Noble Peace prize in 2005. Though Greenpeace denounced the choice because it felt that IAEA had, in fact, increased the risk of nuclear – weapon proliferation[48] the decision of the Noble Committee will boost the moral of the nuclear energy lobby.
- The spoke person of the six member (India, Australia, US, South Korea, Japan and China) Asia Pacific Partnership on Clean Development and Climate, after its first meeting at Sidney in January 2006, commented: "You have to accept that nuclear power plants, civil nuclear power plants are greenhouse friendly."[49]
- After attending the above meeting, the Indian Environment Minister said "We believe that nuclear power should be used in India to promote our emission reduction."[50]

In the dawn of new millennium, taking advantage of the rising petroleum price and global warming debate, nuclear energy is regaining its acceptability in the global energy market. Absence of any viable alternatives energy source has strengthened their position. Unfortunately, decentralized energy sources, like wind, solar and mini hydel project have failed to present themselves as a stable alternative commercial energy source to traditional sources like coal, petroleum and natural gas. It is still playing the role of a supplementary source. In this race for replacing hydrocarbons, for the moment, it is advantage nuclear.

## Signs of Change

Market for better form of energy is also getting ready. In September 2005, the initial public offerings (IPO) of two small German firms ErSol and Q-Cell who make photovoltaic, were oversubscribed many times. During last two years, out of the 13 IPOs registered with the Frankfurt Stock Exchange, four dealt with solar energy.[51] The signs of change are visible now.

Out of compulsion, policy makers may accept nuclear energy at its present form (with high safety and security risk), till the supply of the renewable sources become adequate to meet the increasing energy demands. But world leaders across the globe, as the selective list below indicates, have accepted renewable source as clean and safe 'green' energy of the future. In the joint declaration on renewable energy – 'The Way Forward on Renewable Energy' by the EU at the end of WSSD, Johannesburg, the leaders declared 'We expressed our strong commitment to the promotion of renewable energy and to increase the share of renewable energy sources in the global total primary energy supply. We fully endorse the outcome of the World Summit on Sustainable Development, considering it a good basis for further international co-operation, and intend to go beyond the agreement reached in the area of renewable energy'.[52]

The policy makers have made their commitments very loud and clear. Now, the civil society should unitedly play their constructive role to ensure that all such commitments are implemented much before the date lines. The philosophical strength of the green energy should act as a catalyst which is likely to be stronger than the economic might of its competitors.

Considering the present status of renewable energy, some of the above targets may look ambitious. But if we are passionate about it, these can be achieved within the target date. In their famous book,[53] *Competing for the Future*, the co-author—a famous management strategist—spoke about the 'strategic intent,'[54] which is nothing but an animating dream that energizes an organization in a more positive and sophisticated way. While a strategic architecture may point to the way of future, an ambitious and compelling strategic intent provides the emotional and intellectual energy for the journey. In this case, the passion for greener environment for our children shall act as an 'intellectual and emotional' driving force to achieve these ambitious targets.

For centuries, human civilizations have relied on mother Earth for its energy sources – be it fire woods, coal, petroleum or uranium. It is high time that the Sun and the Wind should take up this responsibility and relieve the ailing Earth.

**Table 3: Renewable Energy Targets**

| Country | Renewable Energy Targets |
|---|---|
| Austria | 78.1% of electricity output by 2010 |
| Brazil | Additional 3300 MW from wind, small hydro, biomass by 2016. |
| Denmark | 29% of electricity output by 2010. |
| Finland | 35% of electricity output by 2010. |
| France | 21% of electricity output by 2010. |
| Greece | 20.1% of electricity output by 2010. |
| Italy | 25% of electricity output by 2010. |
| Portugal | 45.6% of electricity output by 2010. |
| Spain | 29.4% of electricity output by 2010. |
| Sweden | 60% of electricity output by 2010. |
| United Kingdom | 10% of electricity output by 2010. |
| | 75,000 MW by 2010; 22% electricity by 2020. |
| China | The Renewable Energy Law will be effective from 2006. From that year renewable energy will become preferential area for energy development. |

*Source: http/www.iea.org/textbase/pamsdb/grlist.aspx?by=target viewed on 12.11.2005; EIA, 2004 Regional Indicators: European Union (EU).*

## Conclusion

As discussed earlier, renewable energy sources by its very nature, demands proliferation of small-scale mass production units. Retaining centralized control on such technology becomes difficult. Unless ultimate control on the product is fully assured, large corporate houses would not be interested to invest in those projects. In this endeavor for renewable energy, support from mega corporations at this stage is very unlikely. Major initiatives should come from below, at the lowest level – where consumers are really the queens. In addition to the political organizations operating at the grass-root level, only civil societies can organize such a mass movement to save the global climate from the clutches of corporate power brokers.

By shifting environmentally hazardous activities to the fund-starved poorer countries, pollutions will be 'parceled out' to the Southern states. The responsibility of 'green house gas' reduction has been shifted already to the less polluting developing countries by promising few extra dollars. Through the introduction of clean development mechanism, the Kyoto protocol basically has done this. In an integrated global economy, where TNCs have developed alliances with their local partners for further consolidation of their power, an international alliance of consumer and civil societies (which are not nudged into line with grants and assistance from 'foundations' managed by mega corporations) to safeguard common citizens' interest across the globe is essential. Few initiatives to this effect have been taken already. For example, the recent initiative of the Helio International, which is passionately promoting the development and integration of civil society organizations such as users' councils into the energy decision-making process, is a case in point.[55] In the absence of such a coordinated effort, Southern countries, as before, will remain at the receiving ends and would be treated as the dump yards for Northern wastes and pollutants.

## Acronyms

BS: The Business Standard

EPW: The Economic and Political Weekly

ET: The Economic Times

HBL: The Hindu Business Line

ICC: International Chamber of Commerce

IPCC: Intergovernmental Panel on Climate Change

UNCED: The UN.Commission on Environment and Development

UNFCCC: The UN Framework Convention on Climate Change

WBCSD: The World Business Council for Sustainable Development

## Endnotes

1 Isdo, 1992
2 India Today, June 15, 1992
3 OPEC Bulletin, Feb. '92
4 Tanzer, 1974
5 India Today, June 15, 1992
6 Tanzer, 1974
7 The Statesman 13.6.92
8 Dey, 1992
9 Bertelmus, 1999
10 Corporate Europe Observatory (CEO), November 2000, International Group
11 Green peace, August 2002
12 Dhar, 2002
13 Srinivas, 2003
14 EPW, 16.10. 2004
15 Jurewicz, Dawlins, 2005
16 Wirth 2002
17 E.T 5.7.2005
18 E.T 5.7.05
19 Ministry of Non-conventional Energy, 2005
20 HBL. 18/9/05
21 Sington (bbc.co.uk)
22 HBL 21.9.05
23 also see OEF: Is global warming a myth?
24 Riggs, 1998
25 Appenzeller 2004
26 Seshagiri, 1977
27 Dey, 2004
28 Corporate Europe Observatory (CEO), November 2000, April 2001
29 The Businesses Standard, 3.3.2006

30 ET, 3.3.06

31 OPEC, 2004

32 Vanaik, 2005

33 CURES Mailingliste

34 CUTS, 2002

35 Aucland, 2002

36 Afforestation is the direct human-induced conversion of land that has not been forested for a period of at least 50 years to forested land through planting, seeding and/or the human-induced promotion of natural seed sources. Reforestation is the direct human-induced conversion of non-forested land to forested land through planting, seeding or human-induced promotion of natural seed sources, on land that was forested but that has been converted to non-forested land. For the first commitment period (2008-2012), reforestation activities will be limited to reforestation occurring on those lands that did not contain forest on 31st December 1989. Forest is a minimum area of land of 0.05-1.0 hectares with tree crown cover of more than 10-30 percent with trees, with the potential to reach a minimum height of 2-5 metres at maturity in situ. Operational Entity, third-party agency, which certifies the real measurable and long-term emission reductions. Permanence the carbon stocks generated by the project need to be secure over long term and future emission that might arise from this stocks need to be accounted for. Additionally, emission reductions or sequestration must be additional to any that would occur without the project. They must result in a net shortage of carbon and, therefore, a net removal of carbon-di-oxide from the atmosphere.

37 Aucland, 2002

38 E T 10/6/2005

39 Corporate Europe Observer, (CEO) April' 2001

40 Corporate Europe Observatory (CEO), November, 2000

41 Down to Earth, 2005

42 Down to Earth

43 South-South-North (SSN) is a network-based non-profit organization in the fields of climate change and social development that directly pursue structural poverty reduction by building Southern capacity and delivering community based mitigation and adaptation projects. *http://www.southsouthnorth.org/*

44 FORATOM, 2002

45 *NDTV.com*, 21 October 2005

46 Lynn, 2005

47 BS, 30.11.05

48 Koring, 2005

49 CAN-Talk, 2006

50 CAN-Talk, 2006

51 The Economist, October 8, 2005

52 Greenpeace, 2002

53 Hamel, Prahlad, 2002

54 The authors wrote an article on this concept in the Harvard Business Review, 1989, May-June.

55 *http://www.helio-international.org*

## References


1. Appenzeller T, 2004, The Case of the Mission Carbon, National Geographic Magazine, *http://magma.nationalgeographic.com/ngm/0402/feature5/online_extra.html.* (visited on 7/5/05)
2. Aucland L, Moura Costa P, Bass S, Huq S, Landell-Mills N, Tipper R, and Carr R, 2002. Laying the Foundations for Clean Development: Preparing the Land Use Sector: A quick guide to the Clean Development Mechanism. IIED, London. *www.cdmcapacity.org.*
3. Bartelmus P, 1999, Sustainable Development - Paradigm or Paranoia, Wuppertal Papers, Wuppertal Institute for Klima, Umwelt, Energie, Germany.
4. CAN-Talk, 2006, Reuters: INTERVIEW-India says will not agree to emissions caps, can-talk@listi.jpberlin.de (12.1.2006)
5. Corporate Europe Observatory (CEO) (November 2000) Issue Briefing, International Group, Greenhouse Market Mania – UN climate Talks Corrupted by Corporate Psuedo-Solutions, Amsterdam, Netherlands *http://www.corporateeurope.org/greenhouse/internationals.html* (visited on 17.11.05)
6. Corporate Europe Observatory (CEO), November 2000, Issue Briefing – Greenhouse Market Mania, The Climate Fraud Catalogue, *http://www.corporateeurope.org/greenhouse/fraud.html* (visited on 17.11.05)
7. Corporate Europe Observer, (CEO) April 2001, Corporate Campaign to Corrupt the Kyoto Protocol Continues After COP-6, Issue No.8, *http://www.corporateeurope.org/observer8/cop6.html* (visited on 17.11.05)
8. CURES Mailingliste, Report Slams World Bank Role in Clean Energy, *http://listi.jpberlin.de/mailman/listinfo/cures* (mail dated 5.11.2005)

9. CUTS,2002, ODS ( Regulation and Control) Rules, 2000, A Brief Report, September 16, Calcutta.

10. Dey, Dipankar (1992), Of Politics and Global Warming, The Frontier, Kolkata July 25.

11. Dey Dipankar (2004): Anxieties over off- shoring: State vs New Economy, Economic and Political Weekly, Mumbai, June 12.

12. Dhar B, 2002, Private conversation with the author on 'Compatibility of CBD with TRIPS', New Delhi.

13. Down To Earth, 2005, Vol.14 No.12, Nov, 07 *http://www.downtoearth.org.in/fullprint.asp* (visited on 12.11.05)

14. FORATOM (2002), Nuclear energy's contribution to sustainable development, Leaflet distributed at COP 8, New Delhi. *http:// www.foratom.org*

15. Green Peace, August' 2002, *http://archive.greenpeace.org/earthsummit/wbcsd/* (visited on 12/11/05)

16. Greenpeace, 2002, *http://www.greenpeace.org/raw/content/international/press/reports/joint declaration by-eu-other.pdf*

17. *http://www.helio-international.org*

18. Hamel G, Prahalad C K 2002, Competing for the Future, Tata McGraw-Hill Publishing Company Limited, New Delhi.

19. *http://www.ndtv.com/topstories/showtopstory.asp?slug=India%2C+ US+discuss+nuke+coop* (visited on 21.10.05)

20. Isdo S.B (1992), Carbon-di-oxide can revitalize the planet, OPEC Bulletin, March, Geneva.

21. Jurewicz P, Dawlins K, 2005, The Treaty Database – US Compliance with global Treaties, Institute for Agriculture and Trade Policies, Minneapolis, USA. *http://www.nyseg.org/sustain-ed/pages/whtSD/list -text.ht.ml.*

22. Koring P, 2005, ElBaradei, IAEA share Nobel Peace Prize, Globeandmail.com, October 8, *http://www.theglobeandmail.com/servlet/story/LAC.20051008.NOBEL08/BNPPrint/* (visited on 8.10.05)

23. Lynn M, 2005, Nuclear Energy Perks Up on Global Warming, Oil: *Bloomberg.com http://quote.bloomberg.com/apps/news?pid=10000039&refer=ecolumnist_lynn&sid=ahal.s.OT...* (visited on 6/9/05)

24. Ministry of Non-Conventional Energy, 2005, New and Renewable Energy Policy Statement, Draft -II, Govt of India, New Delhi

25. OPEC, 2004, Oil Outlook to 2025, September.

26. Oracle Education Foundation (OEF), Think Quest: Is global warming a myth? *http://library.thinkquest.org/C005858/policy1.html*, visited on 24.3.06

27. Riggs W D, 1998, Global Warming: Divided Science and Unfounded Policy. American Experiment Quarterly, summer.
28. Seshagiri N, 1977, the Weather Weapon, National Book Trust, India.
29. Srinivas N N, (2003), Monsanto Patents Indian Wheat Gene *http://www.indiaresource.org/raw/news/2003/4486.html* (visited on 12.11.05)
30. Sington Da-vid, Global Dimming, Science & Nature: TV & Radio Follow-up, *bbc.co.uk.* Visited on 28.09.06.
31. Tanzer, M, 1974, the Energy Crisis – World Struggle for Power and Wealth, Monthly Review Press, New York.
32. UNEP, 2002, 10 years after Rio: the UNEP assessment, UNEP, Paris
33. Vanaik A, Kyoto Protocol: Tool little, too bad, The Asian Age, 23.3.2005.
34. Wirth Timothy, 2002, Hot air over Kyoto, Environment, Volume-3 (IV) winter.

# ANNEXURE

**Table 1: World Petroleum Consumption 1990-2003 (thousand barrels per day)**

| Region/Country | 1990 | | 2003 | | Growth (%) in 2003 over 1990 Consumption |
|---|---|---|---|---|---|
| | Consumption | % of Share on on total | Consumption | % of Share on total | |
| USA | 16988.5 | 25.52 | 20033.5 | 25.01 | 17.92 |
| France | 1826.1 | 2.74 | 2059.8 | 2.57 | 12.80 |
| Germany | 2681.8 | 4.03 | 2677.4 | 3.34 | (–0.16) |
| Spain | 1010.1 | 1.52 | 1544.3 | 1.93 | 52.89 |
| Sweden | 322.1 | 0.48 | 346.1 | 0.43 | 7.45 |
| United Kingdom | 1776.0 | 2.67 | 1722.4 | 2.15 | (–3.02) |
| Italy | 1873.8 | 2.81 | 1874.4 | 2.34 | - |
| Norway | 200.4 | 0.30 | 257.4 | 0.32 | 28.44 |
| Poland | 282.4 | 0.42 | 476.2 | 0.59 | 68.63 |
| Western Europe | 13306.2 | 19.99 | 14950.6 | 18.67 | 12.36 |
| India | 1168.3 | 1.75 | 2320.0 | 2.90 | 98.58 |
| China | 2296.4 | 3.45 | 5550.0 | 6.93 | 141.68 |
| Japan | 5218.1 | 7.84 | 5578.4 | 6.96 | 6.90 |
| Asia and Oceania | 13718.8 | 20.61 | 22255.5 | 27.78 | 62.23 |
| Eastern Europe and Former USSR | 9731.6 | 14.62 | 5407.9 | 6.75 | (–44.43) |
| **Total World** | **66576.0** | **100.00** | 80098.8 | **100.00** | 20.31 |

*Source: International Energy Annual 2003, Energy Information Administration.*

**Table 2: World Coal Consumption by Region (million short tonnes)**

| Region/Country | 1990 | 2001 | 2025 (Projected) | Average Percent Change Annual 2001-25 |
|---|---|---|---|---|
| North America | 971 | 1148 | 1553 | 1.3 |
| West America | 894 | 574 | 446 | –1.0 |
| Industrialized Asia | 231 | 312 | 400 | 1.0 |
| Total Industrialized | 2095 | 2034 | 2399 | 0.7 |
| East Europe/Former Soviet Union | 1376 | 828 | 649 | –1.0 |
| Developing Countries | 1835 | 2401 | 4433 | 2.6 |
| **World Total** | 5307 | 5263 | 7482 | 1.5 |

*Source: International Energy Outlook 2003, Energy Information Administration.*

3

# Environmental Threat for Developing Economies

*M Ramanjaneyulu*

***The mad rush for rapid economic growth has resulted in degrading the environment. Statistics show that the developing economies are all set to take over the developed countries in causing pollution and disturbing the ecological balance by emitting toxic gases like carbon-di-oxide ($CO_2$). High energy consumption by the developing countries, combined with population explosion and growing needs, is the reason behind the emission of $CO_2$, which has affected the environment. Interestingly enough, the developing countries are themselves the victims of environmental degradation. This paper examines two hypotheses: (1) Carbon-di-oxide emission and its effects on the environment are positively dependent on energy consumption and (2) Energy consumption is increasing due to increase in registration of motor vehicles and industries in the developing countries. The methodology uses a Log-linear model to estimate the carbon-di-oxide emission caused by the developing countries. The author also analyzes various pollutants and their impact on the***

*Source: The Icfai Journal of Environmental Economics, February 2006.* 

***global environment. The data pertaining to global energy consumption in the year 2001 gives a clear picture of the growing demand. The author also discusses the global environmental scenario with special reference to India and its National Environmental Policy, against the backdrop of the Global Environmental Facility Program.***

## Introduction

The importance of environmental issues for the success of efforts for development has been increasing among developing economies across the globe. Environmental degradation can lead to a self-perpetuating process due to economic necessity and communities may inadvertently destroy or exhaust the resources on which they depend for their survival. Rising pressures on environmental resources in a developing world can have severe consequences for self-sufficiency, income distribution, and future growth potential. Environmental degradation can also detract from the velocity of economic development by imposing high costs through expenditure on health and reduced productivity of resources. The poorest 20% of world's population experience the consequences of acute environmental degradation arising out of pressures on marginal land, which has led to falling farm productivity and per capita food production. Similarly, the inaccessibility of sanitation and clean water mainly affects the poor and is believed to be responsible for 80% of the diseases worldwide. The solutions to these environmental problems involve enhancing the productivity of resources and improving the living conditions of the poor and achieving environmentally sustainable development. There are environmental costs and damages associated with the rapid population growth and various expanding economic activities, so environmental consideration is an integral part of policy initiatives for long-term sustainable methods of national productivity. The growing consumption needs of poor countries may have global implications such as destruction of forests, global warming and the greenhouse effect.

## Objectives

The objective of this paper is to examine two things—the environmental devastation which is taking place by energy consumption and the carbon-di-oxide emission which is a potential pollutant for degradation of the environment.

## Hypotheses

The following hypotheses are framed to test the objectives and their effect on Carbon-di-oxide ($CO_2$) emission in the developing countries:

- $CO_2$ emission and its effects on the environment are positively depending on energy consumption.
- Energy consumption is increasing due to increase in registration of motor vehicles and industries in the developing countries.

## Methodology

To estimate the $CO_2$ emission of developing countries, a Log-linear Model has been used for cross-section data for 22 developing nations during 1980, 1990 and 2001. To understand and estimate coefficient of variable energy consumption (in thousand metric tons oil equivalent per country), an independent variable is used for analysis. The extent of influence is estimated by using this model:

$$\ln CDE_i = \beta_1 + \beta_2 \ln \text{ENERGY CON}_{2i} + u_i \qquad \text{.............(1)}$$

In the above equation, $\beta_1$ and $\beta_2$ terms are equal parameters and u equals error term. It is being hypothesized in the model that $CO_2$ emission is positively related to energy consumption. The direct relationship is presumed between the energy use and $CO_2$ emission in developing economies. Environmental damage almost always hits the hardest to those living in poverty. (United Nations, Human Development Report 1998). A new view of a better environment stewardship is essential to sustain development (World Bank, World Bank Atlas, 1997).

**Results:** Estimation of $CO_2$ emission by consumption of energy in developing countries during 1980, 1990 and 2001 is presented in Table 1. It is evident that on an average, one unit of energy consumption led to a 1.238% increase in $CO_2$ emission in 1980, 1.359% in 1990 and 1.344% in 2001. The coefficients of energy consumption (EC) gives the emission elasticity. In practice, however, elasticity is computed at the mean average value of dependent and independent variables. R square of 0.861 means 86% of variation in the logs of $CO_2$ emission which is explained by the logs of energy consumption during 1980; it has slightly increased to 87% in 1990 and retained at 86% by 2001.

**Table 1: Estimation of Carbon-di-oxide Emission by Use of Energy Consumption of Developing Countries**

| Period | Constant | Coefficient (EC) | R | R-Square | Adj. R-Squ. | F | DW |
|---|---|---|---|---|---|---|---|
| 1980 | -3.862<br>(-7.829) | 1.238<br>(11.11) | 0.928 | 0.861 | 0.854 | 123.49 | 1.885 |
| 1990 | -4.477<br>(-8.193) | 1.359<br>(11.34) | 0.931 | 0.867 | 0.860 | 129.94 | 2.211 |
| 2001 | -4.448<br>(-7.797) | 1.344<br>(11.20) | 0.929 | 0.863 | 0.856 | 125.47 | 2.444 |

Note: Figures in the parentheses indicate t-values.

The above results indicate that developing countries are the areas prone to $CO_2$ emission and both energy consumption and emission are functionally related which can be seen through the R square and Durbin Watson Statistics for three decades. Overall R square explains around 86% and DW statistics is also around 2, overall goodness of fit (F-statistics) is highly significant.

## The Global Environment

There is an inverse relationship between the global environmental degradation and growth of world population and income rise. The efficient use of resources and a number of environmental changes will actually provide economic savings, sustainable investments in pollution abatement technology and resource management, as well as significant trade-offs between production and environmental improvements. Issues such as biodiversity, rain forest destruction and population growth will focus international attention on some of the most economically poor countries in the world. The most cumulative environmental destruction has been caused by the developed world. However, high fertility rate, rising average incomes, and increasing inequality in the developing world are some of the factors that cause pollution as they are responsible, directly or indirectly, for toxic emissions and therefore, environmental damage.

### The Environmental Degradation

The most pressing environmental challenges in developing countries in the next few decades will be caused by poverty. These will include health hazards created by lack of access to clean water and sanitation, indoor air pollution from biomass stoves, deforestation and severe soil degradation, all of which are most common

**Table 2**

| Environmental Problems | Effect on Health | Effect on Productivity |
|---|---|---|
| 1. Water Pollution and Water Scarcity | More than 2 million deaths and billions of illnesses a year attribute to pollution; poor household hygiene and added health risks caused by water scarcity. | Declining fisheries, rural household time and municipal costs of providing safe water; aquifer depletion leading to irreversible compaction; constraint on economic activity because of water storages. |
| 2. Air Pollution | Many acute and chronic health impacts; excessive urban particulate matter levels are responsible for 300,000 to 700,000 premature deaths annually and for half of childhood chronic coughing; 400 million to 700 million people, mainly women and children in poor rural areas, affected by smoky indoor air. | Restrictions on vehicle and industrial activity during critical episodes; effect of acid rain on forests and water bodies. |
| 3. Solid and Hazardous Wastes | Diseases spread by rotting garbage and blocked drains; risks from hazardous wastes typically a local but often acute. | Pollution of ground water resources. |
| 4. Soil Degradation | Reduced nutrition for poor farmers on depleted soils; greater susceptibility to drought. | Field productivity losses in the range of 0.5% to 1.5% of gross national product common on tropical soils, offsite siltation of reservoir, river-transport channels, and other hydrologic investments. |
| 5. Deforestation | Localized flooding, leading to death and disease. | Loss of sustainable logging potential and of erosion prevention, watershed stability, and carbon sequestration provided by forests. |
| 6. Loss of Biodiversity | Potential loss of new drugs. | Reduction of ecosystem adaptability and loss of genetic resources. |
| 7. Atmospheric Changes | Possible shifts in vector-borne diseases; risks from climatic natural disasters; diseases attributable to ozone depletion (perhaps 300,000 additional cases of skin cancer per year worldwide; 1.7 mn cases of cataracts.) | Sea rise damage to coastal investments; regional changes in agricultural productivity, disruption of marine food chain. |

where households lack economic alternatives to unsustainable patterns of living. The principal health and productivity consequences of environmental damage in the developing world are divided into water pollution and water scarcity, air pollution, solid and hazardous wastes, soil degradation, deforestation, loss of biodiversity, and atmospheric changes. Table 2 envisages the principal health and productivity consequences of environmental damage in developing countries.

## Pollutants and their Impact on the Global Environment

The rapid increase in the production of pollutants has led to a dramatic increase in the levels of concentration of a number of greenhouses and ozone-depleting gases. For example, the global concentration of $CO_2$ has increased by nearly 30% since the start of the industrial revolution, and more than half of this increase has occurred since 1960. The total gaseous, usually ozone-depleting chlorofluorocarbons (CFCs), increased in concentration by 114% in just 16 years between 1975 and 1990. The level of concentration of another important greenhouse gas, methane, has increased by 143% since the start of the industrial revolution, and almost 30% of this increase has occurred since 1970. Due to rising incomes and rapid population growth, accumulation of these chemicals will accelerate in the future unless sweeping international reforms are implemented. A study jointly sponsored by the World Meteorological Organization and the United Nations Environment Program predicts that if current emission trends continue, the average global temperature is likely to rise 0.3O C per decade or 3O C (5.4 F) by the end of this century. Due to the delayed impact of current emissions, the study found that to stabilize $CO_2$ and CFC concentrations at current levels, immediate reductions in emissions from human activities of over 60% would be required. Most of the warmest years on record have occurred over the past two decades. The evidence may not yet be decisive on whether or not this constitutes a statistically significant pattern of warming even though the trend continued throughout the 1990s, with 1997 and 1998 being the warmest years ever recorded. 2001 was also warmer than 1997, becoming the second hottest year ever.

## Global Energy Consumption

Energy consumption is equal to indigenous production plus imports and stock exchange, minus exports and fuels supplied to ships and aircraft engaged in international transport. Figure 1 shows the global energy use in 2001, which represents high-income countries (HICs) using energy, more than half of the world's energy.

**Figure 1: Global Energy Use 2001**

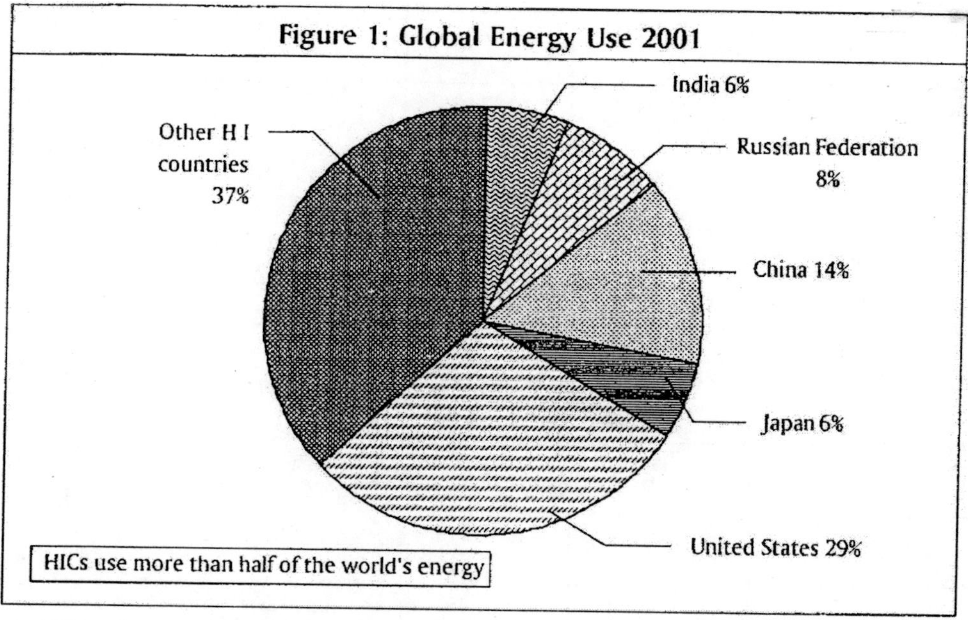


Table 3 represents the increasing decadal growth rate of energy consumption. In absolute terms, energy consumption of all countries has been at an increasing stage. The growth rate of energy consumption in the decade of 1990s is higher

**Table 3: Energy Consumption Per Capita (Total Thousands of Metric Tonnes of Oil Equivalent)**

| Countries | 1980 | 1990 | 2001 | 1980-2001 |
|---|---|---|---|---|
| United States | 1,811,650 | 1,927,572 | 2,281,414 | |
| CGR | | 0.62 | 1.53 | 1.16 |
| China | 598,498 | 870,441 | 1,139,369 | |
| – | | 3.82 | 2.45 | 3.27 |
| Japan | 346,538 | 436,523 | 520,729 | |
| – | | 2.33 | 1.60 | 2.06 |
| **India** | 241,016 | 363,153 | 531,453 | |
| -- | | 4.18 | 3.48 | 4.03 |

*Contd...*

| *Contd...* | | | | |
|---|---|---|---|---|
| Germany<br>– | 360,385 | 356,318<br>–0.11 | 351,092<br>–0.13 | –0.13 |
| France<br>– | 187,737 | 227,114<br>1.92 | 265,570<br>1.42 | 1.75 |
| Brazil<br>– | 111,471 | 132,985<br>1.78 | 185,083<br>3.02 | 2.57 |
| Mexico<br>– | 98,898 | 124,028<br>2.29 | 152,273<br>1.86 | 2.18 |
| United Kingdom<br>– | 201,284 | 212,176<br>0.53 | 235,158<br>0.93 | 0.78 |
| Indonesia<br>– | 59,933 | 92,815<br>4.47 | 152,304<br>4.56 | 4.77 |
| Republic of Korea<br>– | 32,637 | 92,578<br>10.98 | 194,780<br>6.92 | 9.34 |
| Pakistan<br>– | 25,472 | 43,424<br>5.80 | 64,508<br>3.36 | 4.75 |
| Thailand<br>– | 22,162 | 42,215<br>6.66 | 75,542<br>5.37 | 6.32 |
| Malaysia | 12,162 | 22,455<br>6.31 | 51,608<br>7.77 | 7.49 |
| Chile | 9,662 | 13,630<br>3.50 | 23,801<br>5.14 | 4.61 |
| *Source: World Development Indicators 2000 and 2004, Washington DC, New York.* | | | | |

than that of 1920s in all countries. The compound growth rate of energy consumption per capita in the US has increased at 0.62% in the decade of 1980-90, and 1.53% in 1990-2001. Korea Republic has registered the highest growth rate of energy consumption per capita at 9.34% per annum during 1980 and 2001. It was followed by Malaysia with 7.49% and Thailand with the percentage of 6.32%. Energy per capita consumption of Germany has been recording a declining growth.

## Carbon-di-oxide ($CO_2$) Emission

Carbon-di-oxide emissions are stemming from the burning of fossil fuels and manufacturing of cement. They include $CO_2$ produced during consumption of solid, liquid and gas fuels and gas flaring. Table 4 shows the growth rate of $CO_2$ emissions. Large quantities of emissions in absolute quantity of $CO_2$ are produced in developed countries. Fast economic growth is responsible for producing emission in China and India. These countries are responsible for producing at a growth rate of 4.98% per annum, during 1980-90 and 1.51% growth rate of $CO_2$ during 1990-2000. On an average, during 1980-2000, China produced $CO_2$ emission at the rate of 2.32% per annum and India has produced at a growth rate of 5.79% per annum, and low and middle income countries produced at 7.4% per annum. Thus, $CO_2$ emission in developing countries is very high, whereas emissions per unit of production are now lower in developed countries than the middle or low income/less developed countries. Improvements in this type of environmental efficiency of production will have to be accelerated even more. Despite the dramatic improvement in the developed countries' emissions per dollar of GDP, total emissions still increased. Per capita metric tonnes of $CO_2$ emission from China, India, Japan, and low income countries is increasing at a faster rate.

**Table 4: Carbon-di-oxide Emissions**

| Countries | Million Metric Tonnes | | | CGR | Per Capita Metric Tonne | | | Kg per 1995 ppp S of GDP | | |
|---|---|---|---|---|---|---|---|---|---|---|
| | 1980 | 1990 | 2000 | '80-2000 | 1980 | 1990 | 2000 | 1980 | 1990 | 2000 |
| China | 1,476.80 | 2,401.7 | 2,790.5 | | 1.5 | 2.1 | 2.2 | 3.5 | 1,4 | 0.6 |
| CGR | | 4.98 | 1.51 | 3.23 | | | | | | |
| India | 347.3 | 675.3 | 1,070.9 | | 0.5 | 0.8 | 1.1 | 0.7 | 0.5 | 0.5 |
| CGR | | 6.87 | 4.72 | 5.79 | | | | | | |
| Japan | 920.4 | 1,070.7 | 1,184.5 | | 7.9 | 8.7 | 9.3 | 0.8 | 0.4 | 0.4 |
| CGR | | 1.52 | 2.55 | 1.26 | | | | | | |
| Low Income | 774.3 | 1,653.2 | 2,066.7 | | 0.5 | 0.8 | 0.9 | 0.6 | 0.5 | 0.5 |
| CGR | | 7.88 | 2.26 | 5.03 | | | | | | |
| Middle Income | 4,132.90 | 9,169.8 | 9,129.1 | | 2.3 | 3.8 | 3.4 | 1.2 | 1.0 | 0.7 |
| CGR | | 8.30 | –0.04 | 4.04 | | | | | | |
| Low and Middle Income | 2,682.60 | 10,823.2 | 11,196.2 | | 1.5 | 2.5 | 2.2 | 1.0 | 0.9 | 0.6 |
| CGR | | 14.97 | 0.34 | 7.4 | | | | | | |
| High Income | 8,945.60 | 10,480.8 | 11,804.3 | | 12.0 | 11.4 | 12.4 | 1.2 | 0.6 | 0.5 |
| CGR | | 1.6 | 1.2 | 1.4 | | | | | | |
| World | 13,852.70 | 21,297.5 | 22,994.5 | | 3.4 | 4.1 | 3.8 | 1.1 | 0.7 | 0.6 |
| CGR | | 4.4 | 5.2 | 2.56 | | | | | | |

*Source: World Development Indicators 2000 and 2004, Washington DC, New York.*

Developing countries still account for a relatively small proportion of industrial $CO_2$, resulting from the burning of vegetation to clear new land. It is also estimated that deforestation accounts for roughly 25% of all $CO_2$ emissions worldwide. The tropical rainforests represent an important mechanism through which the ecosystem regenerates itself. Cleaning the rain forests will reduce the absorptive capacity of the environment for $CO_2$. The changes in the pattern of land use in developing countries are contributing heavily to global concentrations of greenhouse gases.

## Contributions to Greenhouse Gases

The burning of fossil fuels by automobiles and industries are obvious sources of greenhouse gases; less obvious sources include deforestation, animal husbandry, wet rice cultivation, decomposition of waste, and coal mining. A number of gases, including CFCs, $CO_2$, methane, sulphur dioxide, and nitrous oxides, contribute significantly in the stock of greenhouse gases. However, $CO_2$ has a relatively long lasting influence on the atmosphere and the massive quantities produced globally. Sources of $CO_2$ emissions may be decomposed into two broad categories: Industrial production (77% of emissions) and all others. Developing countries, with roughly three-quarters of the world's population, produce less than one-third of industrial $CO_2$—about one-fifth if we exclude China. Because incomes and consumption are higher in the rich countries, per capita emissions are also much higher. For example, the level of per capita emissions in the US is more than twice that of the average Europeans, 19 times higher than the average African's and 23 times higher than the average Indian's.

## Ozone Depletion

Ozone means a gas composed of three atoms of oxygen. Ozone is a bluish gas that is harmful to breathe. Nearly 90% of the earth's ozone is in the stratosphere and is referred to as the ozone layer. The Ozone layer lies approximately 15-40 km above the earth's surface in the atmosphere. This layer absorbs a band of Ultra Violet radiation called UVB and prevents it from reaching the ground that is particularly harmful to living organisms. Ozone Depleting Substances include CFCs, HCFCs, halons, methyl bromide, carbon tetrachloride and methyl chloroform. Depletion of the ozone layer by ODS will lead to higher UVB levels,

| Sources of Greenhouse Gases | |
|---|---|
| **Emission from Power Plant** | **Methane from Coal mines** |
| • Steel plant<br>• Cement plant<br>• Petroleum refineries<br>• Road transport sector<br>• Lime production<br>• Nitric acid production<br>• Aluminium production<br>• Soda ash use, pulp and paper production<br>• Informal energy, intensive sector: Brick manufacturing and sugar ceramics | • Oil natural gas<br>• Municipal solid waste<br>• Methane emission from waste water (industrial and domestic) |

which in turn will cause increased skin cancers and cataracts and potential damage to marine organisms, plants and plastics.

The increasing evidence regarding the extent of ozone depletion and encroaching global warming presents alarming implications for the global climate. Concerns range from an increase in the incidence of skin cancer to desertification and rising oceans. The cost of climate changes with loss is more for people already living in semi-arid regions. The developing countries have predominantly agricultural economies and so many are located in warm semi-arid regions of the globe; any increase in temperature is likely to have harsh implications on incomes and food self-sufficiency. The land use patterns of developing countries currently make their largest contribution to global concentrations of greenhouse gases. It is essential that deforestation alone accounts for roughly 25% of $CO_2$ emissions worldwide. Deforestation leads to the destruction of a vital source of atmospheric oxygen. Cleaning the rain forests will reduce the environment's absorptive capacity for $CO_2$.

Recently, rapidly increasing global concentrations of atmospheric pollutants have threatened to cause severe damages to the ozone layer as well as dramatic climatic changes such as global warming. To reduce the severity of these environmental threats, global emissions must be sharply curtailed. Table 5 presents atmospheric concentration of greenhouse and ozone depletion during 1980-2001. These constitute carbon tetro chloride, methyl chloroform, CFCs, nitrous oxide and methane, etc. All these are estimated in terms of particles per million and billion. Except Methyl Chloroform, all other gases are at an increasing trend during this period. Therefore, the dust particles of these ingredients in the environment are increasing in the greenhouse which is affecting the health and productivity of the nation.

**Table 5: Atmospheric Concentration of Greenhouse and Ozone Depletion**

| Year | Carbon Tetro Chloride PPM* | Methyl Chloroform PPM** | CFC 11 PPM* | CFC 12 PPM* | CFC 13 PPM* | Total PPM* Chlorine PPM* | Nitrous Oxide PPB** | Methane PPB** |
|---|---|---|---|---|---|---|---|---|
| 1980 | 90 | 71 | 159 | 295 | 25 | 1624 | 301 | 1542 |
| 1981 | 91 | 70 | 166 | 207 | 24 | 1692 | 301 | 1553 |
| 1982 | 93 | 82 | 176 | 324 | 25 | 1965 | 303 | 1562 |
| 1983 | 94 | 86 | 184 | 343 | 28 | 1939 | 304 | 1578 |
| 1984 | 95 | 89 | 192 | 358 | 31 | 2016 | 304 | 1586 |
| 1985 | 97 | 93 | 202 | 379 | 36 | 2121 | 304 | 1604 |
| 1986 | 98 | 97 | 211 | 397 | 40 | 2216 | 305 | 1617 |
| 1987 | 100 | 91 | 222 | 415 | 48 | 2322 | 305 | 1627 |
| 1988 | 101 | 104 | 233 | 436 | 53 | 2425 | 306 | 1635 |
| 1989 | 101 | 109 | 242 | 455 | 60 | 2524 | 307 | 1656 |
| 1990 | 102 | 102 | 251 | 473 | 66 | 2620 | 308 | 1616 |
| 1991 | 102 | 114 | 257 | 487 | 71 | 2685 | 309 | 1673 |
| 1992 | 102 | 118 | 261 | 499 | 79 | 2751 | 310 | 1689 |
| 1993 | 101 | 114 | 263 | 516 | 80 | 2764 | 310 | 1688 |
| 1994 | 101 | 108 | 264 | 511 | 81 | 2769 | 310 | 1693 |
| 1995 | 100 | 99 | 263 | 521 | 82 | 2753 | 311 | 1701 |
| 1996 | 99 | 87 | 263 | 526 | 83 | 2725 | 312 | 1702 |
| 1997 | 98 | 75 | 262 | 531 | 83 | 2693 | 313 | 1710 |
| 1998 | 97 | 64 | 261 | 534 | 83 | 2664 | 313 | 1718 |
| 1999 | 96 | 54 | 260 | 537 | 82 | 2675 | 314 | 1728 |
| 2000 | 95 | 45 | 258 | 539 | 82 | 2695 | 315 | 1729 |
| 2001 | 94 | 40 | 257 | 539 | 81 | 2735 | 315 | 1714 |

Note: *PPM = Parts Per Million; **PPB = Parts Per Billion.

*Source: World Resource Institute (2001), World Resources 2000-01, Washington D.C.*

## Air Pollution in India

The principal sources of air pollution, which pose the greatest health threat associated with modernization, are energy use, vehicular emissions and industrial production. Industrialization can lead to an increase in waste products either through direct emissions or indirectly by the altering patterns of consumption and the boosting demand for manufactured goods. Air pollution consists of particulate matter, sulphur dioxide and nitrogen dioxide. Particulate matter refers to fine suspended particulates less than 10 microns in diameter that are capable of penetrating deep into the respiratory tract and causing severe health damage. The state of the country's technology and pollution controls is an important determinant of particulate matter concentrations. Sulphur dioxide is an air pollutant produced when fossil fuels containing sulphur are burned. It contributes to acid rain which can damage human health, particularly that of the young and the elderly. Nitrogen dioxide is a poisonous, pungent gas formed when nitric acid combines with hydrocarbons and sunlight producing a photochemical reaction. These conditions occur in both natural and anthropogenic activities. Bacteria, motor vehicles, industrial activities, nitrogenous fertilizers, and combustion of fuels and biomass, and aerobic decomposition of organic matter in soils and oceans, which are extremely harmful for health and production, emit nitrogen dioxide. Table 6 represents air pollution in India.

**Table 6: Air Pollution in Indian Cities (Microgram Per Cubic Meter)**

| | Particulate Matter | | Sulphur Dioxide | Nitrogen Dioxide |
|---|---|---|---|---|
| Cities | 1995 | 2001 | 1995-2001 | 2000-2001 |
| Ahmedabad | 289 | 104 | 30 | 21 |
| Bangalore | 123 | 56 | – | – |
| Kolkata | 375 | 154 | 49 | 34 |
| Chennai | 130 | 125 | 15 | 17 |
| Delhi | 415 | 187 | 24 | 41 |
| Hyderabad | 152 | 51 | 12 | 17 |
| Kanpur | 459 | 136 | 15 | 14 |
| Lucknow | 463 | 136 | 26 | 25 |
| Mumbai | 240 | 79 | 33 | 39 |
| Nagpur | 185 | 69 | 6 | 13 |
| Pune | 208 | 58 | – | – |

*Source: World Development Indicators 2000 and 2004, Washington DC, New York.*

## Rain Forest Preservation

Preservation of forests in the developing world depends on traditional fuels. The vast majority of wood cut is used for domestic purposes such as home heating and cooking. Timber accounts for 80% of the total fuel in Africa, 70% in South America, and 74% in Asia. In order to halt the destruction of rain forests permanently, the respective problems must be addressed. Long-term solutions will involve increasing the accessibility of cheap alternative fuels, managing sustainable timber schemes, and providing economic opportunities for the masses of impoverished people now resorting to clearing large tracts of fragile rain forest land. The efficiency of the existing rain forest use can be improved. Most of the timber that is now burned to open land for cultivation could be harvested for financial gain. For example, it is estimated that Brazil loses $2.5 bn annually in the burning of precious rain forest timber. Sustainable timber production for fuel or export can be achieved through the restriction of cutting cycles to 30-year intervals, and careful maintenance of the new growth. Another important boon to forest conservation would be the development of markets for alternative rain forests products. The costs of rain forest preservation can be compensated by developing markets for sustainable forest products such as meats, nuts, fruits, oils, sweeteners, raisins, tannins, fibers, construction materials, and medicinal compounds may provide a more lucrative and sustainable stream of income from tropical forests. Yet, only less than 0.1% of tropical forests are sustainably managed. Forests can be saved through appropriate aid packages, which alleviate landlessness and poverty, and in turn, help to eliminate the socioeconomic causes of tropical deforestation.

Deforestation leads to permanent conversion of natural forests area to other uses and includes shifting of cultivation, agriculture, ranching, settlements, and infrastructure development. The deforested area and areas degraded by fuel wood gathering, acid rainfall, and forest fires are also intended for regeneration. Average annual deforestation is more in low-income countries in the world as shown in Table 7. High income countries could be able to protect almost 20% of the total area designated as scientific reserves and natural reserves for wildlife sanctuaries, protected landscape and seascapes with limited public access.

**Table 7: Deforestation and Biodiversity**

| Countries | % of Forest Area to Total Land Area | | Average Annual Deforestation | | % of Nationally Protected Area to Total | |
|---|---|---|---|---|---|---|
| | 1995 | 2000 | 1990-96 | 1990-2000 | 1996 | 2003 |
| China | 17.4 | 17.0 | 0.1 | -0.9 | 6.4 | 7.8 |
| India | 21.9 | 21.6 | 0.0 | -0.1 | 4.8 | 5.2 |
| Japan | 66.8 | 66.1 | 0.1 | 0.0 | 6.8 | 3.4 |
| Low income | 17.8 | 27.1 | 0.7 | 0.8 | 5.9 | 8.4 |
| Middle income | 32.7 | 32.7 | 0.3 | 0.1 | 4.9 | 9.1 |
| Low and middle income | 26.5 | 30.7 | 0.4 | 0.3 | 5.3 | 8.9 |
| High income | 20.8 | 26.1 | -0.2 | -0.1 | 10.5 | 19.5 |

*Source: World Development Indicators 2000 and 2004, Washington DC, New York.*

## Global Environmental Facility Program

Figure 2 reveals the funds for global environmental facility programs region wise. Allocation of funds for the Global Environment Facility Program February 1991-January 2004 amounted to $19,944 mn. Asia attained the highest share of the facility at 41%. Latin America and Africa have received to the extent of 21% and 20% of funds for the facility program.

**Figure 2: Funds for Global Environmental Facility Programs by Region**

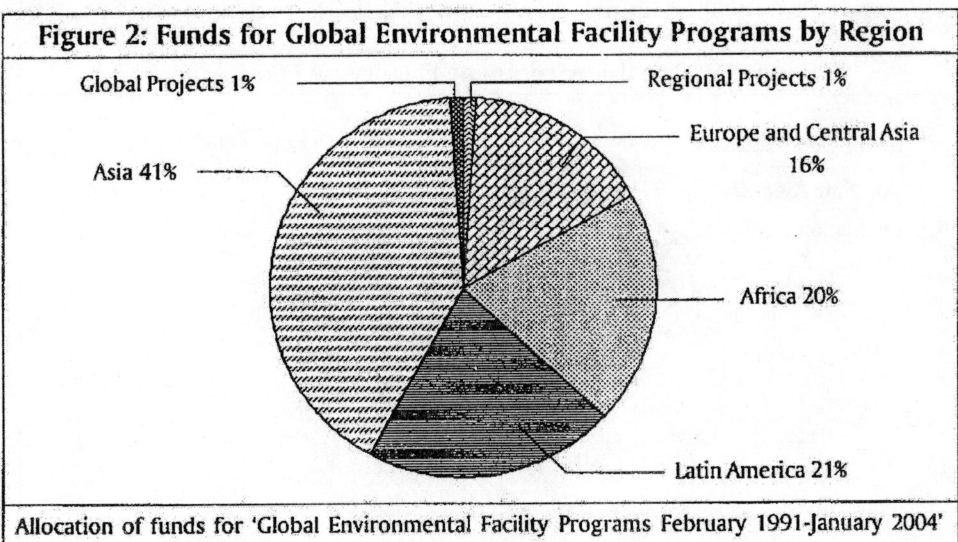


Allocation of funds for 'Global Environmental Facility Programs February 1991-January 2004' amounted to $19,944 mn.

The fund facility for crucial area such as climate change and biodiversity has received the share of 54% and 26% respectively. Correspondingly, international water, multiple areas and ozone depletion have also received very low attention facility i.e., 11%, 6% and 2%.

## Pursuit for Solution on Biodiversity and Pollution

The first United Nations Conference on Environment and Development (UNCED), the so-called Earth Summit, was held in Stockholm in 1972. The second Summit took place in Rio de Janeiro in June 1992. The Rio meeting brought together the leaders of 118 industrial and developing countries along with hundreds of non-governmental organizations and tens of thousands of concerned individuals. The main task was to find ways to manage the increasing dangers of permanent environment damage, resulting from the build-up of greenhouse gases (especially $CO_2$), leading to fears of global warming, the inexorable loss of biodiversity (diverse plants and species) resulting in part from the destruction of tropical rain forests, and concerns over the environmental consequences of rapid population and industrial growth in the developing world.

The emphasis of the Rio Summit was on allocating development assistance to programs on poverty alleviation and environmental health such as sanitation, clean

**Figure 3: Global Environment Facility by Focal Area**

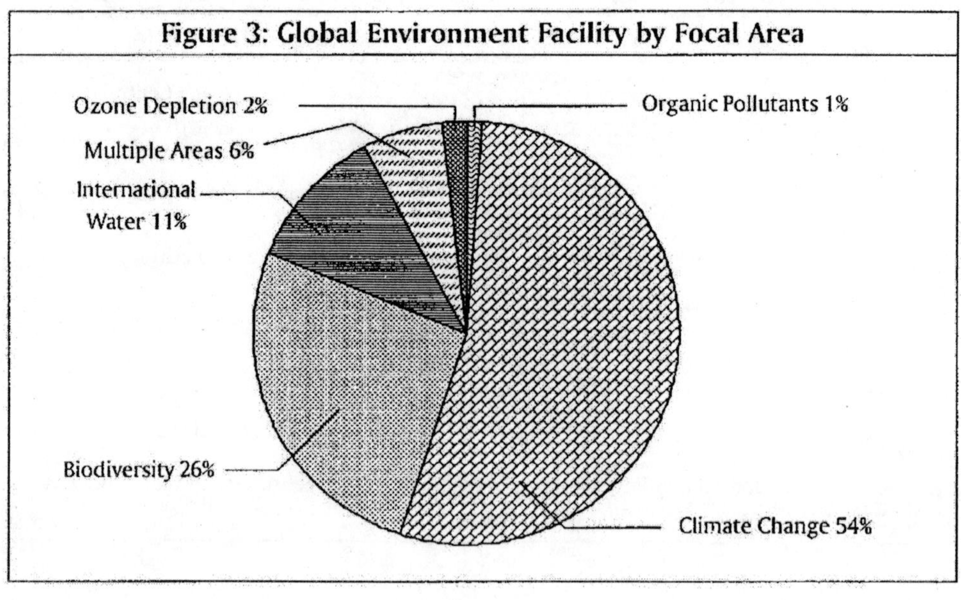

water, reducing air pollution from burning of firewood, and meeting the basic needs. Investment in research and extension services to reduce soil erosion permits more environmental-sensitive agricultural practices. Allocation of more resources is also needed to reduce population growth. Supporting the governments of developing countries in their attempt to curtail or modify projects that harm the environment is very essential. Provision of funds is essential to protect the natural habitats and biodiversity of a region. Investing in research and development on alternative energy is a step to respond to climate changes and reduce greenhouse gases.

In November 2001, representatives of 180 countries agreed on methods to enforce the Kyoto accord. The agreement calls on the industrial countries to reduce greenhouse gases, especially $CO_2$ from industry and cars. Each country must reduce emissions by an average of 5.2% below 1990 levels by 2012. Analysts predict that total global emissions of $CO_2$ will still increase by some 1.5 billion tonnes per year by 2010 and global warming could become dangerous as early as 2050.

## National Environmental Policy

The development of international conventions about the importance of the long-term survival of ecosystems and ultimately humanity is a major motivator for including global issues in the national agenda. Financial resources and technology is needed to solve environmental problems and improve local technical knowledge. The environmental standards at the production of products and manufacturing should be a part of trade restrictions. The consumer attention should be on environmental friendliness of products. Sustainable development is possible through responsibility of markets and governments to control prices, mainstreaming environmental considerations and developing sustainability indicators. Economic instruments are used to direct environmental policy. The economic approach may be simply stated in a set of propositions. Environmental degradation has an economic cost viz., natural capital stock, natural forest, GNP, aesthetic damage. Population growth, insecure resource rights, bad central planning, the failure of markets to price resources and outputs also reflects social costs.

## Conclusion

The developing countries should formulate appropriate environmental policy, which is crucial for the management of resources. Majority of global environmental damages represent the cumulative and contemporary impact of Western industrialization. Developing countries have the potential for environmental degradation but ultimately lead to the balance against human survival. The developed world was successful with economic growth but the heavily populated developing countries could bring negative environmental externalities to the world and then the rich and poor would suffer alike. The adjustment programs of IMF and World Bank stabilization policies ensured that the continued poverty of the nations in the South is a necessity for maintaining the attained prosperity of the nations in the North. The environmental interdependence is allowed to developing countries to get financial assistance to pursue environmentally sustainable development objectives.

*(M Ramanjaneyulu is a Professor, PG Department of Economics, Bangalore University, Karnataka. He can be reached at metta_ramanjaneyulu@yahoo.com)*

## References

1. David W Pearce and Jeremy J Warford (1993), *World without End: Economics, Environment, and Sustainable Development—A Summary,* Washington DC, World Bank.
2. Government of India (2001), *Central Statistical Organization,* Compendium of Environment Statistics, New Delhi.
3. Government of India (2004), Ministry of Environment and Forest, "India's Initial National Communication to the United Nations Framework Convention on Climate Change", New Delhi.
4. Jose, I, Dos R, Furtado and Tanara Belt, Ramachandra Jammi (2000), World Bank Indicators, Economic Development and Environmental Sustainability Policies and Principles for a Durable Equilibrium, Washington DC, USA.
5. Michael P Todaro, Stephen C Smith (2004), "Economic Development", Pearson Education, Delhi.
6. Sous La Director De Denizhon Eroccal (1991), *Environment Management in Developing Countries,* OECD, France.
7. World Development Indicators (2004), World Bank, Washington DC.
8. World Development Report (2001), World Bank, Washington DC.
9. World Resource Institute (1998), *World Resources: A Guide to the Global Environment,* Oxford University Press, New York.

# 4

# Environmental Crime in Global Context
## Exploring the Theoretical and Empirical Complexities

*Rob White*

***This paper aims to explore the conceptual and research challenges of studying environmental harm as a criminological phenomenon. The paper discusses issues relating to scale (e.g., local, regional, global) and impacts (e.g., immediate, long term), as these relate to matters pertaining to definition and jurisdiction (e.g., international law and law enforcement, eco-human rights). In addition to mapping out complexities involved in the examination of environmental harm, the paper attempts to identify potential directions for further theoretical reflection and empirical research.***

## Introduction

Environmental issues have generated considerable public interest in recent years, and not surprisingly, criminologists and other social scientists are now likewise turning their attention to how best to define and respond to environmental harm (Lynch and Stretsky, 2003; White, 2003). In so far as major environmental

*Source: http://www.criminology.law.usyd.edu.au/. This article was earlier published in Current Issues in Criminal Justice, Vol. 16, No. 3, pp. 271-285.* 

changes are occurring on the global scale, with significant impacts at the local level, so too greater urgency and critical analysis about environmental matters has grown. Simultaneously, similar kinds of local issues are being repeated across the globe, making us realize that the global and the local are frequently intertwined and in many ways inseparable. This is often encapsulated in the term 'glocalization' (see Crowley, 1998).

The task of trying to understand, interpret and act upon matters that are often systemic, complicated and intrinsically interconnected poses certain dilemmas for the criminologist. For instance, our interest and knowledge in this area may well be growing (albeit from a rudimentary base), but the more we know, the less secure we seem to be in the knowledge that we have. The very complexities of the issues can make it daunting to tackle them. It certainly makes things analytically challenging.

Consider, for example, the following observations. The development of a green or environmental criminology as a field of sustained research and scholarship will by its very nature, incorporate many different perspectives and strategic emphases. Environmental criminology deals with concerns across a wide range of environments (e.g., land, air, water) and issues (e.g., fishing, pollution, toxic waste). It involves conceptual analysis as well as practical intervention on many fronts, and includes multi-disciplinary strategic assessment (e.g., economic, legal, social and ecological evaluations). It involves the undertaking of organizational analysis, as well as investigation of 'best practice' methods of monitoring, assessment, enforcement and education regarding environmental protection and regulation. Analysis needs to be conscious of local, regional, national and global domains and how activities in each of these overlap. It likewise requires cognisance of the direct and indirect, and immediate and long-term, impacts and consequences of environmentally sensitive social practices (White, 2003: 484).

One challenge for environmental criminology is to separate out different levels and kinds of analysis, and to 'make sense' of what is a very complicated whole. This is the intent of the present paper. That is, I wish to explore the conceptual and research challenges of studying environmental harm as a criminological phenomenon. In order to do so, I wish to utilize an analytical mapping exercise that covers key areas of potential interest to criminologists. This is folio wed by an appraisal of how environmental crime itself is socially constructed. To illustrate this, we will consider issues pertaining to fishing and social regulation.

## Analytical Mapping of Environmental Harm

To understand complexity, we need to simplify. My objective in this section is to identify some important areas for analytical consideration and to discuss these in abstract conceptual terms. Specifically, I wish to discuss environmental issues in regard to four types of perspective: Focal considerations; geographical considerations; locational considerations; and temporal considerations (see Figure 1).

**Figure 1: Mapping of Environmental Harm**

| | | | |
|---|---|---|---|
| *Focal Considerations:* | | | |
| [Identify issues pertaining to victims of harm] | | | |
| Environmental Justice | | Ecological Justice | |
| [human beings] | | [biosphere, including plants and animals] | |
| *Geographical Considerations:* | | | |
| [Identify issues pertaining to each geographical level] | | | |
| International | National | Regional/State | Local |
| *Locational Considerations:* | | | |
| [Identify issues pertaining to specific kinds of sites] | | | |
| 'Built' Environments | | 'Natural' Environments | |
| [e.g., urban, rural, suburban] | | [e.g., ocean, wilderness, desert] | |
| *Temporal Considerations:* | | | |
| [Identify issues pertaining to changes over time] | | | |
| Environmental Effects | Environmental Impact | Social Impact | |
| [short-term/long-term] | [manifest/latent] | [immediate/lasting] | |

Exploration of themes and issues within each of these areas expose the diversity of perspectives, approaches and concepts that are utilized in the areas of environmental criminology.

## Focal Considerations

By 'focal' considerations, I refer to concerns that center on the key actors or players who are at the center of investigation into environmental harm. In other words, the emphasis is on identifying issues pertaining to the victims of harm, including how to define whom or what is indeed an environmental 'victim'.

How we understand the relationship between human beings and the environment is crucial to defining and responding to environmental issues (see Figure 2). Different perspectives or eco-philosophies include: Anthropocentric (or human-centred); biocentric (or species-centred); and ecocentric (socio-ecological centred). These perspectives can be assessed on the basis of how they conceive environmental problems, how they depict the role of humans in the production of such problems and how they approach the issue of environmental regulation (see Halsey and White, 1998).

**Figure 2: Eco-Philosophies**

**Anthropocentrism**

- Superiority of humans over all other living and non-living entities
- Non-human nature viewed instrumentally
- Humans are separate from world's ecosystem
- Ideological basis for human activity is self-interest
- Highly centralized organs of power: Nation-states and corporations
- Concern with economic interests and profits
- Strategy of 'sustainable development'
- Emphasis on 'management' of resources
- Domination of humans by humans, and domination of non-human nature by humans.

**Biocentrism**

- Biocentric equality
- Human beings have same moral worth as other 'species' on planet
- Non-human species have intrinsic value
- Ideological basis for activity is preservation and realization of all species
- Reduces the social to the biological: 'survival of the fittest'
- 'Natural selection' of human beings (via war, disease, famine) is not a problem
- Human pursuits should ideally be directed by an understanding of 'Gaian Truth'
- Decisions concerning the environment should be made according to which outcomes are most likely to foster the widest possible diversity of life, both non-human and human
- Emphasis on 'righteous management' involving mass preservation of wilderness etc.
- Issue of interplay between human rights and 'biotic' rights.

*Contd...*

| *Contd...* |
|---|
| **Ecocentrism** |
| • Humans and their activities are inextricably integrated with the rest of the natural world |
| • Humans have the capacity to deploy methods of production which have global consequences; therefore, they have a responsibility to ensure that such production methods do not exceed the ecospheric limits of the planet |
| • Live simply so that others [human and non-human] may simply live |
| • Balance between instrumental and intrinsic conceptions of non-human nature |
| • Dialectical nature of the relationship between human action and non-human processes, interconnectedness of life |
| • Social justice is equally important and inextricably bound to issues of ecology |
| • Work with non-human nature, and a commitment to collective needs |
| • Principles of participatory democracy, via bioregionalism, and constant movement between local initiative and global solidarity |
| *Adapted from Halsey and White, 1998.* |

For many of those working on environmental issues, the question of broad philosophy translates into specific concerns with the idea of eco-human rights or ecological citizenship (see for example, Halsey, 1997; Smith, 1998). What does this mean in practice? It means that present generations ought to act in ways that do not jeopardize the existence and quality of life of future generations. It also means that we ought to extend the moral community to include non-human nature. By doing so, we enter a new politics of obligation:

> *"In ecological thought, human beings have obligations to animals, trees, mountains, oceans, and other members of the biotic community. This means that human beings have to exercise extreme caution before embarking upon any project which is likely to have the possibility of adverse effects upon the ecosystems concerned (Smith, 1998:99)".*

This particular notion of ecological citizenship thus centres on human obligations to all living things, and to carefully assess the impacts of human activity across the human and non-humans domains.

However, such considerations are not without their problems. Thus, the conceptualization of 'rights' is itself contentious when extended to the non-human (see Christoff, 2000). For example, should environmental rights be seen as an extension of human or social rights (e.g., related to the quality of human life, such as provision of clean water), or should human rights be seen as merely one component of complex ecosystems that should be preserved for their own sake (i.e., as in the notion of the rights of the environment)? While increasingly acknowledged in international law, the environment connection with human rights continues to be somewhat ambiguous and subject to diverse practical interpretations (Thornton and Tromans, 1999). Nevertheless, such ambiguities and tensions over 'rights' are essential parts of the criminological debates characteristic of the shift from eco-philosophy to conceptions of environmental crime.

Within criminology, there are significant issues surrounding scale, activities and legalities as these pertain to environmental harm. A strict legalist approach tends to focus on the central place of criminal law in the definition of criminality. Thus, as Situ and Emmons (2000: 3) see it: 'An environmental crime is an unauthorized act or omission that violates the law and, is therefore, subject to criminal prosecution and criminal sanctions'. However, other writers argue that, as with criminology in general, the concept of 'harm' ought to encapsulate those activities that may be legal and 'legitimate' but which nevertheless negatively impact on people and environments (Lynch and Stretsky, 2003).

The responses of the state to environmental harm (however defined) are guided by a concern with environmental protection. This is generally framed in terms of ensuring future resource exploitation, and dealing with specific instances of victimization that have been socially defined as a problem. Risk management in this case is directed at preventing or minimizing certain destructive or injurious practices into the future, based upon analysis and responses to harms identified in the present.

Analysis of environmental issues proceeds on the basis that someone or something is indeed being harmed. *Environmental justice* refers to the distribution of environments among peoples in terms of access to and use of specific natural resources in defined geographical areas, and the impacts of particular social practices and environmental hazards on specific populations (e.g., as defined on the basis of class, occupation, gender, age, ethnicity). In other words, the concern is with human beings as the center of analysis. The focus of analysis, therefore, is on human health and well-being and how these are affected by particular types of production and consumption.

Here we can distinguish between environmental issues that affect everyone, and those that disproportionately affect specific individuals and groups (see Williams, 1996; Low and Gleeson, 1998). In some instances, there may be a basic 'equality of victims', in that some environmental problems threaten everyone in the same way, as in the case, for example, of ozone depletion, global warming, air pollution and acid rain (Beck, 1996). As extensive work on specific incidents and patterns of victimization demonstrate, however, it is also the case that some people are more likely to be disadvantaged by environmental problems than others. For instance, American studies have identified disparities involving many different types of environmental hazards that adversely affect people of color throughout the United States (Bullard, 1994). There are thus patterns of 'differential victimization' that are evident with respect to the siting of toxic waste dumps, extreme air pollution, access to safe clean drinking water, etc. (see Chunn *et al.* 2002; Williams, 1996). Another dimension of differential victimization relates to the subjective disposition and consciousness of the people involved. The specific groups who experience environmental problems may not always describe or see the issues in strictly environmental terms. This may be related to knowledge of the environmental harm, explanations for calamity and socioeconomic pressures to 'accept' environmental risk (see Julian, 2004). The environmental justice discourse challenges the dominant discourses by placing inequalities in the distribution of environmental quality at the top of the environmental agenda (see Julian, 2004; Harvey, 1996).

By way of contrast, *ecological justice* refers to the relationship of human beings generally to the rest of the natural world, and includes concerns relating to the health of the biosphere, and more specifically plants and creatures that also inhabit the biosphere (see Benton, 1998; Franklin, 1999). The main concern is with the quality of the planetary environment (that is frequently seen to possess its own intrinsic value) and the rights of other species (particularly animals) to life free from torture, abuse and destruction of habitat. Specific practices, and choices, in how humans interact with particular environments present immediate and potential risks to everything within them.

In specific areas, concepts such as speciesism may be invoked. This refers to the practice of discriminating against non-human animals because they are perceived as inferior to the human species in much the same way that sexism and racism involve prejudice and discrimination against women and people of different color (Munro, 2004). However, it is important to recognize that the environmental justice discourse is critical of many mainstream environmental groups precisely because of their 'focus on the fate of "nature" rather than humans' (Harvey, 1996: 386). To put it differently, taking action on environmental issues involve choices and priorities. Many communities who suffer from the 'hard end' of environmental harm feel that their well-being ought to take priority over 'natural environments' or specific plants and animals as such.

## Geographical Considerations

Students of environmental harm have to be cognizant of the varying issues that pertain to different geographical levels. As alluded to above, some issues are of a planetary scale (e.g., global warming), others regional (e.g., oceans and fisheries), some are national in geographical location (e.g., droughts in Australia), while others are local (e.g., specific oil spills). Similarly, laws tend to be formulated in particular geographically defined jurisdictions. With regard to nation-states such as Australia, relevant laws include international law, federal laws, state laws and local government bylaws.

Intervention on environmental issues requires not only new concepts of justice and rights, they also require acknowledgment of transnational processes and responsibilities. It has been pointed out that:

> *"...transnational economic processes, transcontinental cultural links and transboundary environmental impacts have generated a new democratic deficit -the remedy of which requires new forms and institutions for democratic participation which extend beyond the borders of the nation-state (Christoff, 2000: 200)".*

The telecommunications revolution has brought the world into the lounge rooms of the advanced industrialized countries and extended the scope of our knowledge of the fate of previously unheard of places and species. It has also expanded public or common sense knowledge of the interconnected nature of environmental processes (and harms), which finds expression in the catchphrase 'Think globally, act locally'. Institutionally, the concern with environmental well-being is manifest explicitly in the priority areas of international policing.

From the point of view of international law enforcement agencies, the major issues relating to environmental crime are:

- The trans-border movement and dumping of waste products;
- The illegal traffic in real or purported radioactive or nuclear substances; and
- The illegal traffic in species of wild flora and fauna.

These areas have been identified by agencies such as Interpol as key subjects in relation to environmental crime. It is worth exploring the first of these in greater depth, given that much of the transfer of waste has been from advanced industrialized countries to 'third world' countries.

The biggest exporter of toxic waste is the United States. Hazardous residues and contaminated sludge are most likely to find a foreign home in a Third World country. The pressures for this are twofold. On the one hand, the US has seen the closing of many domestic landfills due to public health problems, and increasing public consciousness of the dangers posed by toxic waste. On the other hand, poor countries (and corrupt state officials) may find it financially attractive to offer their land as sites for US waste (see Rosoff, Pontell and Tillman, 1998).

The problem is not only the transfer of toxic waste; it is the generation of toxic waste in other countries by companies based in advanced industrialized nations. The classic case of this are the *maquiladoras,* American-owned factories set up across the border in Mexico. Here, environmental regulation is lax, resulting in high levels of chemical pollution, contamination and exposure to toxic materials. Closer to home for Australians, is the huge environmental damage caused to the *Ok Tedi* river in Papua New Guinea by the activities of the Australian mining corporation BHP (see Low and Gleeson, 1998). Because the PNG government was dependent on the earnings from the *Ok Tedi* copper mine, it actively cooperated with BHP in the destruction of local rain forest and much of the river system. Many villagers have lost the entire environment that supported their way of life (Low and Gleeson, 1998: 8).

These examples highlight the fact that to understand the overall direction of environmental issues demand analysis of the strategic location and activities of transnational capital, as supported by hegemonic nation-states on a world scale. Capitalist globalization, bolstered via neo-liberal state policy, means that there is great scope to increase environmentally destructive activity. This is demonstrated in how the traffic in risk occurs at the global level where developing countries play the same role as the poorer communities within the developed nations (e.g., 'business-friendly' countries that accept hazardous industries and toxic wastes). The issue here is how best to respond to NIMBY (Not In My Backyard) opposition within developed countries in ways that do not simply shift the problem elsewhere.

> *"The structural difference of economic needs and government regulation between the developed and developing worlds, and the absence of any supranational body to ensure consistency in environmental standards, has encouraged western industrial capital to shift unpopular and increasingly illegal hazard-producing activities and wastes across national boundaries to states which often define, and welcome, these transfers as 'investment' (Low and Gleeson, 1998: 121)".*

For criminologists, the challenge is to incorporate notions of environmental justice into their overall analytical framework by maintaining a sense of global scale. It also requires understanding of the political economy of environmental harm (White, 2002).

## Locational Considerations

We can make a distinction between geographical area and 'place'. The latter refers to specific kinds of sites as described in the language of 'natural' and 'built' environment. There is considerable overlap, interconnection and interplay between these types of environments. Nevertheless, the distinction is useful, particularly when assessing which environmental issues appeal to which sections of the population and for what reasons (Tranter, 2004).

In simple terms, we can describe the 'Built' environment as basically referring to significant sites of human habitation and residency. It includes urban and rural areas, and areas of crossover between the two consisting of major regional concentrations of people, commuter suburbs and zones, etc. The 'Natural' environment consists of wilderness, oceans, rivers and deserts. These are sites in which human beings may be present, or through which they may traverse, but which are often seen as distinctive and 'separate' from human settlement *per se* (however, this needs to be qualified by acknowledging different ways in which humans interact with their environments, reflecting different cultural and material relationships to the land – see Langton, 1998).

What constitutes an environmental harm or environmental crime is partly a matter of visibility of the issues, partly a matter of public policy. What can be identified via personal experiences, expert representation or sectional interest group as being worthy of attention, is that which is most likely to gain recognition as a public issue (see Hannigan, 1995). Meanwhile, governments have laws across a wide range of issues, relating to air, water, toxic waste, use of public lands, endangered species and the list goes on. The relationship between public policy and government strategic action is also shaped by contingency—specific events, situations and disasters tend to shake things up rapidly and with immediate effect.

**Figure 3: Coloring Environmental Issues**

**'Brown' issues**

- Air pollution
- Pollution of urban stormwater
- Pollution of beaches
- Pesticides
- Oil spills
- Pollution of water catchments
- Disposal of toxic/hazardous waste

**'Green' Issues**

- Acid rain
- Habitat destruction
- Loss of wildlife
- Logging of forests
- Depletion of ozone layer
- Toxic algae
- Invasive species via human transport
- Water pollution

**'White' Issues**

- Genetically modified organisms
- Food irradiation
- In vitro processes
- Cloning of human tissue
- Genetic discrimination
- Environmentally-related communicable diseases
- Pathological indoor environments

The precise nature of an environmental issue is in itself linked to specific group interests and consciousness of harm. For example, environmental issues have been categorized according to three different types of harm (Crook and Pakulski, 1995; Tranter, 2004; see also Curson and Clark, 2004). These are set out in Figure 3.

The significance of conceptualizing environmental issues in this way is that it demonstrates the link between environmental action (usually involving distinct types of community and environmental groups), and particular sites (such as urban centers, wilderness areas or seacoast regions). Some issues tend to resonate more with members of the public than others; other issues generally only emerge if an accident or disaster brings it to the fore.

The mobilization of opinion is crucial to determination of what is or is not considered a 'crime' (or 'harm'), and how the state will, in the end, respond to the phenomenon in question. The complex relationship between human and non-human 'rights' is thus played out in practice through the importance of 'place' in the lives of diverse communities. This inevitably leads to conflicts over purposes, as each place or site is subject to competing demands-jobs (via logging), recreation (via tourism), sustenance (via settlement), aesthetics (via photography) etc. Disputes over value and use are settled using the full range of political, ideological, legal, coercive and persuasive means available to stakeholder parties.

## Temporal Considerations

Another key issue for consideration relates to issues pertaining to changes over time. To some extent, such considerations are ingrained in contemporary environmental impact assessment in the guise of the 'precautionary principle' (Harvey, 1998; Deville and Harding, 1997). That is, what we do with and in the environment has consequences, some of which we cannot foresee.

Temporal considerations can be distinguished in terms of environmental effects, environmental impacts and social impacts. The short-term effects of environmental degradation include such things as the release of chlorofluorocarbons into the atmosphere, the long-term effect being the accumulation of greenhouse gases and ultimately climate warming. Environmental impacts begin with global warming as a manifest consequence of planetary change, and results in the latent consequences of changes in sea levels and changes in regional temperatures and precipitation (among other things). The social impact of environmental change are both immediate, as in the case of respiratory problems or increased probability of disease outbreak, and long-term (e.g., lower quality of life, alteration of physiological functioning).

From the point of view of eco-philosophy, the tendency has been for anthropocentric perspectives to dominate when it comes to answering the questions, *what to do, over what period of time?* And yet, protection of the environment very often requires criteria that go beyond a human-centred approach. To put it differently, the appropriate time scale for understanding resource and population stability is generally much longer than we are used to:

> Different systems move along different time scales. Geology works in the millions of years; economics in the tens of years; biology from a few minutes to a few centuries; evolutionary biology from a few years to millions of years. Appropriate time scales depend on how long it takes for things to happen in the subject area (Page, 1991: 64).

The importance of temporal concerns is reflected in cultures that view the relationship between people and the environment in holistic, reciprocal terms. The concept of 'balance' in some indigenous communities, for example, reminds of vital significance: 'This precarious balance still exists, and the relationship between plants, animals, the elements, the air, water, wind and earth are all equally and evenly placed within the whole' (Robyn, 2002: 202). Here, we see a value system and code of ethics that embodies living within one's means and living within and as part of nature (see also Langton, 1998). It is an ecocentric approach to life.

The philosophy of living in and with nature is empirically reflected in two phenomena: One relating to 'place', the other to 'time'.

> The diversity of Native cultures and kinds of social organizations which developed through time represent a high degree of social/political complexity and are varied according to the demands and necessities of the environment.

For example, American Indian nations organized at the band level of social/political development have used effective strategies to take advantage of marginal habitats such as the Arctic and deserts of the Americas where resources are limited' (Robyn, 2002: 198-199).

Importantly, such systems are usually decentralized, communal and self-reliant: 'These societies live closely with and depend on the life contained in that particular ecosystem. This way of living enabled indigenous communities to live for thousands of years in continuous sustainability' (Robyn, 2002: 1999).

The point of this discussion is that evaluation of environmental issues needs to consider the element of time: Negatively, from the perspective of short and long-term consequences of environmental harm; positively, from the perspective of 'what works' in protecting and preserving environments.

In summary, I have tried to demonstrate that there are a number of intersecting dimensions that need to be considered in any analysis of specific instances of environmental crime. These include consideration of who the victim is (human or non-human); where the harm is manifest (global through to local levels); the main site in which the harm is apparent (built or natural environment); and the time frame within which harm can be analyzed (immediate and delayed consequences). While this represents a form of analytical mapping of environmental harm, that illustrates the complexities of such analysis, it remains to be seen how such mapping might assist in explaining the 'real world' of environmental criminalization.

## Social Construction of Environmental Issues

To some extent, an abstract model or mapping of environmental harm can be useful in exposing areas of further research and consideration, beyond that dealt with formally by law enforcement agencies and the criminal justice system at present. However, it can also be used to assist in explaining why it is that some types of human activity are more likely to be subject to criminalization than others. The theme of this section is how environmental crime is socially constructed. Specifically, the concern is to identify those elements that together result in activity being deemed harmful, and thereby worthy of investigation and prosecution.

When considering these matters, it is useful to bear in mind the following questions (see White, 2004):

### What is the Problem?

In order to do this we have to deal with issues of definition and evidence of harm. We have to analyze potentially competing claims as to whether or not the problem exists, and diverse lay and expert opinion on how the problem is interpreted. Does it pose a risk, and if so, to whom, and in what ways? Is the initial problem serious enough in the public's eye to warrant a social response in the form of community action or state intervention?

### Why does the Problem Occur?

To answer this we need to examine the social context, and to investigate the actions of key actors involved with the phenomenon.

### What are the Social Dynamics that Allow the Problem to Persist or Ensure that State Action is Taken to Overcome it?

To answer this, we need to tackle issues pertaining to the shaping of perceptions, interpretation of events, and intervention processes. Is the problem socially constructed as a social problem warranting social action? In what ways is the problem construed from the point of view of social regulation and what forms of state and private intervention are mobilized to contain or manage the problem? Is the problem itself to be addressed, or is the focus on how best to avoid, cover-up or manage any risk associated with the problem?

### A Case Study: Abalone Theft

In recent years, the stealing of abalone has come to prominence and, indeed, is touted as one of the key areas in which environmental crime, as crime, is being addressed in a concerted way in Australia. We want to know why this is the case, especially given that environmental harm in many other cases draws much less state attention.

The abalone industry is highly regulated, with strict quotas enforced, limited numbers of licensed divers and extensive documentation of each catch required. Part of the reason for this high level of regulation is that the industry is a major export earner, bringing in over $100 mn a year. Australia produces about one-third of the global wild abalone harvest, and it has been pointed out that 'Australia's stake in global supply has increased following the decline and/or disappearance of abalone populations in other parts of the world—including Japan, Mexico, South Africa and the United States (California)—due to negative environmental conditions, limited stocks, illegal fishing and poor fisheries management' (Taiby and Gant, 2002: 1). Global demand for abalone, and high profits from abalone sales, have contributed to the growth in illegal harvesting.

The illegal abalone market has been described in terms of five categories of offender (Tailby and Gant, 2002). In summary, these include:

- Organized poachers who operate in crews and harvest large quantities
- Licensed divers who engage in over-quota fishing and docket fraud
- Shore-based divers who access certain poaching spots
- Extended family groups who engage in double-bagging
- Individuals who take over-bag limit.

Our main interest here is with the organized stealers of abalone (although there is some overlap with licensed divers, who may use the same networks for processing and distributing the catch). There are several features of these groups that warranted particular attention (Tailby and Gant, 2002; Leonard, 2004; Little, 2004). For example, organized poachers frequently have sophisticated infrastructure to facilitate the theft boats, infrared night vision equipment, scuba gear, hired transport vehicles, light aircraft etc. Illegal processing of the abalone may also be quite sophisticated, and involve canning, drying or cryovac (vacuum) packaging.

Abalone thieves of this kind are willing to cross state borders to harvest abalone. Increasingly, it appears that organized criminal groups are moving into the industry, including outlaw motorcycle gangs and Asian crime figures. The illicit networks extend across state boundaries (from Tasmania to Queensland, or Victoria to New South Wales, for example). They also cross international boundaries, as one of the more lucrative markets for illegally harvested abalone is Asia. It has also been suggested that there are links between trade in illegal Australian abalone and the illicit drug markets. Again, these links transcend state and national boundaries.

Given the negative impact of illegal harvesting, use and sale of abalone on the legitimate industry, on royalty/tax revenue to the state and on abalone stocks generally, concerted efforts have been made to counter the illegal industry. Illegal accessing and processing of abalone is criminalized, both in terms of the law and in terms of resources put into the law enforcement process. Thus, 'Each abalone-producing state has legislation carrying high pecuniary penalties and custodial sentences for abalone offending, and has dedicated abalone-crime investigators' (Tailby and Gant, 2002: 5). In Tasmania, for example, offenders may be prosecuted under the state's Criminal Code for offences such as lying to public officials and receiving or possessing stolen property, or they may be subject to two indictable offences under the Living Marine Resources Management Act, 1995 that refer to illegally taken fish and falsifying documents (Leonard, 2004; Little, 2004). Each area of law imputes that the illegal action is treated as a serious matter. This is also apparent in the penalties assigned to offenders. For example, as a result of the joint efforts of the National Crime Authority and Tasmania Police in 'Operation Oakum', an investigation into abalone theft, several people have been sentenced to prison, including a two-year term of imprisonment in one particular case (Australian Crime Commission, 2004; see also Tasmania Police, 2004).

Investigation of abalone-related criminality features the use of a broad spectrum of police powers, including phone taps, dedicated surveillance, monitoring of documentation, and surprise inspections of processing facilities (Little, 2004; Leonard, 2004; Tailby and Gant, 2002). The cross-border elements of the crime mean that it is of interest and concern to national law enforcement agencies such as the National Crime Authority (now the Australian Crime Commission), to state police services, to relevant fisheries bodies both at the national (National Fisheries Compliance Committee) and state levels (e.g., Fisheries Monitoring and Quota Audit Unit, Tasmania), to the Australian Customs Service, and the Australian Quarantine Inspection Service. In other words, dealing with the crime necessarily involves a wide range of agencies at the local, regional, national and international levels. Cooperation amongst enforcement and monitoring agencies is essential, and agencies such as the NCA have played an important role in providing cross-jurisdictional coordination, access to substantial investigatory powers and use of advanced surveillance technologies.

There are a number of interrelated reasons why abalone theft has been defined and successfully prosecuted as an environmental crime. The social construction of environmental harm, in this instance, is largely due to the complexities of the issues and, ultimately, the economic bottom line (see Chart 4).

Analysis of the different dimensions of an environmental issue can be used to explain both why some activities are subject to criminalization, and why some are not. A case study approach can provide useful insights into how and why this is so.

**Chart 4: Mapping of Abalone Stealing as a Crime**

| | |
|---|---|
| *Focal Considerations:* | |
| [Identify issues pertaining to victims of harm] | |
| **Criminalization** | Link to breaches of criminal law; criminal breaches of maritime law. |
| **Anthropocentric** | Link to business interests, state income and exploitation of resource for human benefit. |
| *Geographical Considerations:* | |
| [Identify issues pertaining to each geographical level] | |
| **Cross-border** | Link to national/state mobilization of resources, facilities and powers. |
| **International** | Operational matters, international trade. |
| *Locational Considerations:* | |
| [Identify issues pertaining to specific kinds of sites] | |
| **Ocean/Coastal** | Link to nature of detection/surveillance [crime initiated]. |
| **City/Factories** | Link to surveillance, use of telephones, communications, transportation [crime realized]. |
| *Temporal Considerations:* | |
| [Identify issues pertaining to changes over time] | |
| **Resource depletion** | Link to immediate and longer-term impacts, especially in the light of world share of market. |
| **Realization of value in formal and informal markets** | Link to current global price for abalone. |

## Conclusion

The intention of this paper has been to acknowledge the different ways in which environmental crime can be analyzed, and to identify potential dimensions that might be considered in investigations of environmental issues. The paper has discussed issues relating to focal targets (e.g., human beings and non-human nature), scale (e.g., local, regional, global), consequences (e.g., immediate, long-term), and time-scale in gauging events, activities and harms. A series of charts were introduced as a means to illustrate key points and to summarize potential areas of analytical concern. By charting out the issues in this way, we are better able to map future research projects and to gauge what further work needs to be done to advance environmental criminology as a field of inquiry.

*(Rob White is a Professor and Head at School of Sociology and Social Work, University of Tasmania. He can be reached at R.D.White@utas.edu.au).*

## References

1. Australian Crime Commission (2004) Annual Report 2002-03. Canberra: ACC.
2: Beck, U (1996) "World Risk Society as Cosmopolitan Society? Ecological Questions in a Framework of Manufactured Uncertainties", *Theory, Culture, Society,* 13(4): 1-32.
3. Benton, T (1998) "Rights and Justice on a Shared Planet: More Rights or New Relations?", *Theoretical Criminology,* 2(2): 149-175.
4. Bullard, R (1994) *Unequal Protection: Environmental Justice and Communities of Color.* San Francisco: *Sierra Club Books.*
5. Christoff, P (2000) "Environmental Citizenship", in W Hudson and J Kane (eds) *Rethinking Australian Citizenship.* Melbourne: Cambridge University Press.
6. Chunn, D, Boyd, S and Menzies, R (2002) '"We All Live in Bhopal": Criminology Discovers Environmental Crime', in Boyd, Chunn and Menzies(eds) *Toxic Criminology: Environment, Law and the State in Canada.* Halifax: Fernwood Publishing.
7. Crook, S and Pakulski, J (1995) "Shades of Green: Public Opinion on Environmental Issues in Australia", *Australian Journal of Political Science,* 30: 39-55.
8. Crowley, K (1998) "'Glocalisation' and Ecological Modernity: Challenges for Local Environmental Governance in Australia", *Local Environment,* 3(1): 91-97.
9. Curson, P and Clark, L (2004) "Pathological Environments", in R White (ed) *Controversies in Environmental Sociology.* Melbourne: Cambridge University Press.
10. Deville, A and Harding, R (1997) *Applying the Precautionary Principle. Sydney:* The Federation Press.
11. Franklin, A (1999) *Animals and Modern Cultures: A Sociology of Human-Animal Relations in Modernity.* London: Sage.

12. Halsey, M (1997) "Environmental Crime: Towards An Eco-Human Rights Approach", *Current Issues in Criminal Justice,* 8(3): 217-242.

13. Halsey, M and White, R (1998) "Crime, Ecophilosophy and Environmental Harm", *Theoretical Criminology,* 2(3): 345-371.

14. Hannigan, J (1995) *Environmental Sociology: A Social Constructionist Perspective.* London: Routledge.

15. Harvey, D (1996) *Justice, Nature and the Geography of Difference.* Oxford: Blackwell.

16. Harvey, N (1998) *Environmental Impact Assessment: Procedures, Practice, and Prospects in Australia.* Melbourne: Oxford University Press.

17. Julian, R (2004) "Inequality, Social Differences and Environmental Resources", in R White (eds) *Controversies in Environmental Sociology.* Melbourne: Cambridge University Press.

18. Langton, M (1998) *Burning Questions: Emerging environmental issues for indigenous peoples in northern Australia.* Darwin: Centre for Indigenous Natural and Cultural Resource Management.

19. Leonard, Gary (2004) Personal communication, November 9, 2004 [in his capacity as Chief Investigations Officer, Fisheries Monitoring and Quota Audit Unit, Tasmania].

20. Little, Colin (2004) Personal communication, November 9, 2004 [in his capacity as Tasmania Police member of Operation Oakum].

21. Low, N and Gleeson, B (1998) *Justice, Society and Nature: An Exploration of Political Ecology.* London: Routledge.

22. Lynch, M and Stretsky, P (2003) "The Meaning of Green: Contrasting Criminological Perspectives", *Theoretical Criminology,* 7(2): 217-238.

23. Munro, L (2004) "Animals, 'Nature' and Human Interests", in R White (eds), *Controversies in Environmental Sociology.* Melbourne: Cambridge University Press.

24. Page, T (1991) "Sustainability and the Problem of Valuation", in R Costanza (eds) *Ecological Economics: The Science and Management of Sustainability.* New York: Columbia University Press.

25. Robyn, L (2002) "Indigenous Knowledge and Technology", *American Indian Quarterly,* 26(2): 198-220. Rosoff, S, Pontell, H and Tillman, R (1998) *Profit without Honor: White-Collar Crime and the Looting of America.* Upper Saddle River, USA: Prentice Hall.

27. Situ, Y and Emmons, D (2000) "Environmental Crime: The Criminal Justice System's Role in Protecting the Environment" *Thousand Oaks:* Sage.

28. Smith, M (1998) *Ecologism: Towards Ecological Citizenship.* Minneapolis: University of Minnesota Press.

29. Tailby, R and Gant, F (2002) "The Illegal Market in Australian Abalone", *Trends and Issues in Crime and Criminal Justice* No.225. Canberra: Australian Institute of Criminology.

30. Tasmania Police (2004) Annual Report 2002-03. Hobart: Department of Police and Public Safety.

31. Thornton, J and Tromans, S (1999) "Human Rights and Environmental Wrongs: Incorporating the European Convention on Human Rights: Some Thoughts on the Consequences for UK Environmental Law", *Journal of Environmental Law,* 11(1): 35-57.

32. Tranter, B (2004) "The Environment Movement: Where to from Here?", in R White (eds) *Controversies in Environmental Sociology.* Melbourne: Cambridge University Press.

33. White, R (2002) "Environmental Harm and the Political Economy of Consumption", *Social Justice,* 29(1-2): 82-102.

34. White, R (2003) "Environmental Issues and the Criminological Imagination", *Theoretical Criminology,* 7(4): 483-506.

35. White, R (2004) "Criminology, Social Regulation and Environmental Harm" in R White (ed) *Controversies in Environmental Sociology.* Melbourne: Cambridge University Press.

36. Williams, C (1996) "An Environmental Victimology", *Social Justice,* 23(4): 16-40.

# Section II

# Government Initiatives

# 5

# National Environmental Policy 2006: Some Observations*

*Usha K R*

***This paper attempts to bring out the salient features of the National Environment Policy (NEP) 2006 released by the Ministry of Environment and Forests. It highlights the imperative need to maintain balance between economic development and environmental protection and focuses on the key environmental challenges that the country faces, relate to the nexus of environmental degradation with poverty in its many dimensions, and economic growth. Research on feasibility of social forestry, provision of safe drinking water, strict enforcement of the rules relating to the water, air and sound pollutions are suggested as the timely measures to safeguard the environment from certain threats. The dominant theme of this policy is that while conservation of environmental resources is necessary to secure livelihoods and well-being of all, the most secure basis for conservation is to ensure that people dependent on particular resources obtain better livelihoods from the fact of conservation, than from degradation of the resource.***

* This article draws upon the National Environment Policy 2006, Government of India, Ministry of Environment and Forests.

## 1. Introduction

Environmental issues have to be considered in the context of the economic, social and political milieu of a country. This is even truer for a diverse, developing nation like India, which faces multiple challenges of alleviation of mass poverty and population growth, infrastructural development, providing livelihood security, healthcare, education, empowerment of the underprivileged and elimination of disparities such as caste and gender.

The issues of Environmental Management in the Indian context have been addressed extensively through policies such as the *National Forest Policy, 1988*, the *National Conservation Strategy and Policy Statement on Environment and Development, 1992*, the *Policy Statement on Abatement of Pollution, 1992*, the *National Agriculture Policy, 2000*, the *National Population Policy, 2000* and the *National Water Policy, 2002*. All these policies have articulated the need for sustainable development in their domain areas and recommended strategies and action plans to achieve it. The National Environment Policy (NEP) is a further step in that direction, supplementing areas of lacunae in previous policies and building on their experience.

NEP has been evolved and carefully designed to address the need for sustainable development which ensures the enhancement of the quality of life for human beings, through methods which conserve natural resources and environment. It works towards social justice for the marginalized sections of society whose lives and livelihoods are threatened by industrialization and unplanned urbanization which can lead to environmental degradation. It acknowledges the organic link between natural resources and the communities dependent on particular resources for their livelihood. It stresses that development should go hand in hand with conservation of natural resources and at the same time, lead to better livelihood for those dependent on a particular resource. It is also an expression of India's commitment to international movements towards environmental conservation.

The policy has been evolved through an intensive and extensive process of consultation with Government bodies at all levels, experts, representatives of major academic and research institutions, key industry associations, voluntary organizations, and eminent personalities in the field of environmental conservation. It is intended to provide guidelines for regulatory and legislative reform, and environmental conservation programs and projects by the Central, State, and Local Governments and other institutions for incorporating environmental concerns in all developmental activities.

## 2. Objectives

The main objectives of the policy are based on current perceptions of major environmental challenges and are open to revision and evolution over time. They are:

### 2.1 Conservation of Critical Environmental Resources

To conserve and preserve natural resources, man-made heritage and environmental systems which are mandatory for livelihoods, economic growth, and human welfare.

### 2.2 Intra-Generational Equity: Livelihood Security for the Poor

To guarantee fair access to environmental resources and value for all communities, specifically the deprived communities, who rely mostly on ecological resources for their well-being.

### 2.3 Inter-Generational Equity

To make sure that application of ecological reserves is intended towards satisfying the wants and ambitions of both the current and future generations.

### 2.4 Integration of Environmental Concerns in Economic and Social Development

To integrate environmental concerns into all economic and social development activities and programs.

### 2.5 Efficiency in Environmental Resource Use

To achieve efficient use of environmental resources to minimize the harmful impact on the environment.

### 2.5 Environmental Governance

To apply the principles of good governance, such as participation, accountability, transparency, using cost effective measures and rationality, to the management of use of environmental resources.

### 2.6 Enhancement of Resources for Environmental Conservation

To augment the surge of sources like social capital, management skills, technology, finance and traditional knowledge towards ecological protection by mutually beneficial partnerships amidst private institutions, academicians, researchers, local communities, public agencies and investors.

## 3. Principles

The policy is based on the conviction that development is sustainable only when it respects ecological constraints and the need for social justice. It aims at eliciting the involvement of public authorities at the Central, State and Local Government levels in crucial strategic interventions to realize its objectives. It is intended to stimulate and facilitate partnerships between such agencies, and evolve legislation to help achieve its objectives. Central to the policy are a set of principles that act as guidelines for all the parties involved in achieving the objectives of the policy. They are:

### 3.1 Human Beings are at the Center of Sustainable Development Concerns

Concerns for sustainable development stem from a primary concern for human beings who are entitled to a healthy and productive life in harmony with nature.

### 3.2 The Right to Development

The right to development must be fulfilled to meet developmental needs of present and future generations along with their environmental needs.

### 3.3 Environmental Protection is an Integral Part of the Development Process

Sustainable development can be achieved only by ensuring that environmental protection is an integral part of the development process.

### 3.4 Precautionary Approach

"Better to err on the side of caution" should be the approach adopted in cases of credible threat to important environmental resources.

### 3.5 Economic Efficiency

There will be a concerted effort to ensure economic efficiency in various public actions for environmental conservation. This means that there will be an attempt to maximize welfare for all members of society through natural and man-made resources. This principle also implies that polluters compensate for the adverse impact on others by bearing the cost, or benefit, of the results of their actions. Efficiency of resource use can also be minimized by minimizing wasteful use and consumption of natural resources, and streamlining processes and procedures to minimize cost and delay.

## 3.6 Entities with "Incomparable" Values

Significant risks to human health, life, environmental life-support systems, and unique natural and man-made entities like historical monuments such as the Taj Mahal, are termed "Incomparable" as damage to these entities cannot be compensated for in terms of money or conventional goods and services. Allocation of resources for the protection and conservation of such entities would have to be prioritized, irrespective of the direct economic benefit derived from it.

## 3.7 Equity

Equality of treatment and justice in terms of allocation of entitlements, obligations and outcomes and participation in processes of decision-making, overuse of environmental resources, must be provided for all sections of society, irrespective of disparities such as caste, gender or generation.

## 3.8 Legal Liability

The current environmental redressal mechanism which is mostly based on criminal liability doctrine has not proved worthwhile and, therefore the, mechanism has to be complemented suitably. Civil charges on environmental damages would daunt the ecologically detrimental acts and therefore, recompense the casualties suitably. Both acts of commission and omission related to an environmental standard would have to be compensated for. As in the instance of the Supreme Court's decisions in the Shriram Gas Leak case and the Bhopal Gas Leak case, strict liability applies whenever the liable party damages a third party.

## 3.8 Decentralization

Decentralization can be accomplished by re-assigning the authority from a Central to State and Local, so as to empower the public authorities to tackle specific environmental issues in their local context.

### 3.9 Integration

Integration must be realized by including environmental considerations in formulation of policies, social and natural sciences in environment related research, and strengthening connections between the Central, State, and Local Self-Government bodies and various agencies concerned with the implementation of environmental policies.

### 3.10 Environmental Standard Setting

Environmental standards must reflect and be appropriate for the society or context in which they apply as different standards are suitable for different social contexts.

### 3.11 Preventive Action

Preventing environmental damage from occurring is better than attempting to restore degraded environmental resources.

### 3.12 Environmental Offsetting

Threatened or endangered species and natural systems that are important for sustaining life, providing livelihoods, or general well-being, have to be protected. In specific instances where in the interest of the larger good such protection cannot be provided, the affected public must be duly compensated for their loss.

## 4. Strategies and Actions

The objectives and principles articulated in the policy are to be realized through concrete action in specific contexts. Such action is ongoing and has been pursued for many years. To add to this body of work, based on the current context, fresh action plans have to be prepared and implemented by all the concerned agencies in order to realize the objectives of the policy. Government bodies at all levels—Central, State and Local, would be encouraged and empowered to formulate their own strategies and actions in keeping with the policy and a mechanism for ensuring their involvement will be institutionalized to achieve the policy's objective of integration of environmental concerns in all developmental processes and policies. The list of strategies and actions articulated in the policy covers ongoing activities and provides guidelines for future action. The list is illustrative and not exhaustive.

## 5. Regulatory Reforms

Inadequacies in legal framework and regulatory institutions for environmental conservation have led to accelerated environmental degradation, delayed action and high costs of transaction in development projects. At present, the basic enactments relating to environment management are *Environment Protection Act 1986, Water (Prevention and Control of Pollution) Act, 1974, Water Cess Act, 1977* and *Air (Prevention and Control of Pollution) Act, 1981, Indian Forest Act, 1927, Forest (Conservation) Act 1980, Wild Life (Protection) Act, 1972B* and *Bio-diversity Act, 2002.* These and other regulations complement these enactments. To ensure that this body of existing legislation and policies is consistent with the objectives of the NEP, the steps taken will be institutionalizing a holistic and integrated approach to environment management consistent with the policy, identifying and integrating environmental concerns in relevant policies, identifying emerging areas for new legislation, encouraging and facilitating review of legislation at the State level, adopting and institutionalizing techniques for assessment of policies and programs to address potential impact and enhance potential favorable impact, and ensuring accountability of all levels of Government in undertaking action to meet the policy objectives, especially with regard to ensuring the livelihoods and well being of the poor through improved access to natural resources.

### 5.1 Process Related Reforms

The existing procedures for granting clearances and other approvals will be reviewed with the aim of reducing delays and levels of decision making, achieving decentralization of environmental processes and achieving greater transparency and accountability. The NEP mainly follows the recommendations of the *Committee on Reforming Investment Approval and Implementation Procedures* (The Govindarajan Committee) which identified delays in environment and forest clearances as the largest source of delays in development projects. These include the *Environment Protection Act, Forest Conservation Act, the Water (Prevention and Control of Pollution) Act, the Air (Prevention and Control of Pollution) Act, the Wildlife (Protection) Act, and Genetic Engineering Approval Committee (GEAC) Rules under the Environment Protection Act.* Besides this, faster decision making, greater transparency and access to information will be realized by promoting use of information technology tools and training related to their use. Greater decentralization will be achieved by empowering and encouraging independent action and regulatory decision making on the part of Government bodies at all levels along with increased accountability. Regulatory mechanisms and processes will be put in place to ensure adherence to principles of good governance.

### 5.2 Framework for Legal Action

At present environmentally harmful behavior is penalized through criminal laws which has largely proved more detrimental, whereas civil law offers flexibility and its sanctions can be adapted to particular situations. Thus, a judicious mix of civil and criminal legislation is the best approach to deal with environmentally harmful behavior. In extreme cases criminal law can be resorted to.

### 5.3 Substantive Reforms

Substantive reforms are related to environment and forest clearances, coastal areas and resources, LMOs (Living Modified Organisms), environmentally sensitive zones, monitoring of compliance, use of economic principles in environmental decision making, pollution abatement, enhancement and conservation of environmental resources, and preserving biodiversity and ecosystems.

## 6. Environment and Forest Clearances

New projects would be appraised and reviewed mainly based on the Environmental Impact Assessment (EIA), with major revisions in the process of assessment in keeping with the recommendations of the Govindarajan Committee. These revisions call for marked decentralization of power so that State/UT level Government bodies become more effective. This has to be supported with adequate human and institutional capacity building. To make the forest clearances processes more effective, assessment of new projects are to be regularized and institutionalized at all levels of Government so that potential hazards to the environment are identified at the planning stage itself. The quality and productivity of land to be diverted for development activities are to be appraised and diversion of such land, especially dense forests, are to be minimized. Further measures would be to encourage and facilitate participatory processes involving agencies at all levels and the potentially affected community, and formulate and regularly review principles of good governance for environmental management in all development activities.

## 7. Coastal Areas

Development activities in the coastal areas follow the guidelines in the Coastal Regulation Zone notifications and Integrated Coastal Zone Management (ICZM) plans. But the Coastal Regulation Zone notifications need to be reviewed to incorporate a holistic approach to coastal environmental regulation, thus providing effective protection of valuable coastal environmental resources without affecting livelihoods, legitimate coastal economic activity, infrastructure development, settlements and coastal ecosystems. The Integrated Coastal Zone Management (ICZM) plans need to be comprehensive, and prepared on a sound scientific basis by experts with the participation of local communities, both in formulation and implementation.

## 8. Living Modified Organisms (LMOs)

While biotechnology has great potential to provide livelihoods and contribute to economic growth, LMOs may pose significant risks to ecological resources and the health of human beings and animals. To ensure that development of biotechnology does not prove hazardous, periodical reviews of regulatory processes of LMOs and National Bio-Safety guidelines will be conducted to ensure that they are based on current scientific knowledge, take ecological, health and economic concerns into consideration, and ensure conservation of biodiversity and human health.

## 9. Environmentally Sensitive Zones

Environmentally sensitive zones are those areas with identified environmental resources of "Incomparitive Value" which need special conservation efforts. To conserve and enhance these resources, without affecting legitimate development activities in these areas, such areas will be identified and granted legal status, and development plans will be formulated and implemented in consultation with local communities.

### 9 1 Monitoring of Compliance

Adequate monitoring and enforcement of compliance with environmental policies and standards will be ensured by facilitating private-public partnerships and empowering and building capacities of local bodies such as Panchayati Raj institutions and urban local bodies to effect monitoring of compliance with environment management plans.

## 9.2 Use of Economic Principles in Environmental Decision Making

Systems of natural resource accounting must be put in place to track enhancement or depletion of such assets. This would also provide a mechanism to make users of resources accountable for their use of resources and its impact on the environment. This will be achieved by strengthening the initiatives taken by the Central Statistical Organization towards natural resource accounting, developing and promoting the use of standardized practices and norms in environmental accounting and the creation of a National Environment Restoration Fund from user fees for access to resources and donations to be used for restoration of environmental resources.

## 9.3 Enhancing and Conserving Environmental Resources

The causes of degradation of natural resources range from environmentally detrimental production and consumption practices and shortcomings in policy making and programs to lack of awareness of the causes and effects of environmental degradation and how to prevent them. Enhancing and conserving environmental resources involves preventing and mitigating land degradation, preserving desert ecosystems, forests and wildlife, biodiversity, traditional knowledge and natural heritage, and conserving fresh water resources such as rivers, groundwater and wetlands, enhancing mountain ecosystems and coastal resources, abating pollution, preserving man made heritage sites and dealing with global climate change.

Degradation of land, through soil erosion, alkali-salinization, water logging, pollution, and reduction in organic matter content must be countered by creating awareness and encouraging adoption of scientific and traditional sustainable land use practices, promoting reclamation of wasteland and degraded forestland through partnerships between investors, land owners and local communities, incorporating strategies for watershed management, reversing desertification and expanding green cover, and encouraging agro-forestry, organic farming, environmentally sustainable cropping patterns, and adoption of efficient irrigation techniques.

The Indian desert area, which constitutes 38.8% of India's land area and is spread over 10 states, is extremely rich in diversity of migratory birds and mammals. But the rapidly increasing population on this area has led to an urgent need for measures to conserve these desert ecosystems. Desert ecosystems must be conserved and enhanced by intensive water and moisture conservation through scientific and traditional practices, expanding green cover, reviewing agronomical practices in these areas and promoting practices that are well adapted to the desert ecosystem.

It is well known that forests are invaluable for environmental conservation in many ways. They recharge mountain streams that form rivers, conserve soil, prevent floods and drought, bring rain, provide homes for wildlife and natural growth of diverse flora and fauna, provide homes and livelihoods for communities dependent on them and yield fuel and other wood. Forests have great potential for economic growth especially ecotourism. It is, hence, imperative that forest cover be restored and conserved through formulating strategies for increase of forest and tree cover and involving large scale participation of various agencies such as Van Panchayats, local communities and local government bodies. Women must be involved as much as possible in management of resources as they are most affected by resource degradation.

Conservation of wildlife is to be ensured by expanding the PA (Protected Area) network in the country and Conservation and Community Reserves, reviewing norms and criteria for placing particular species in different schedules of the Wildlife Protection Act and formulating and implementing strategies for conservation of endangered species outside protected areas while reducing scope for man-animal conflict.

India is rich in both genetic and scientifically developed diversity of flora and fauna, as well as ethno-biology knowledge. This enormous resource base has great potential for economic growth. A National Biodiversity Strategy and Action Plan is being formulated through large scale collection of inputs. Besides this mammoth effort, conservation and enhancement of these resources would involve strengthening the protection of areas rich in biodiversity, ensuring minimization of potential impact of development projects on biodiversity, and harmonizing the provisions of The Patents Act, 1970 to enable local communities holding knowledge of use of resources to benefit from disclosure of that knowledge.

Siltation and sediment in freshwater sources like rivers have to be mitigated by promoting integrated approaches to management of river basins by river authorities, preventing and reducing pollution loads, involving local communities in afforestation of banks and catchments to prevent soil erosion and increase green cover.

Groundwater pollution and lowering of water tables must be countered by encouraging practices such as drip irrigation and rainwater harvesting, formulating and implementing mechanisms for regulation of ground water in industrial and commercial establishments, taking remedial measures to prevent flow of toxic waste into ground water and making the use of pollutants such as fertilizers and pesticides optional.

Wetlands are an invaluable source of water and need to be conserved through formulating strategies for prudent use, ecotourism, protecting specific wetlands by granting them the status of entities of incomparable value, integrating wetland conservation in developmental activities and promoting traditional practices and techniques for conserving village ponds and tanks.

Mountain ecosystems provide forest cover, feed river systems, enrich biodiversity and provide livelihood and economic growth through their potential for ecotourism. Conservation of mountain ecosystems calls for measures towards appropriate land use planning and watershed management practices for their sustainable development, ensuring minimization of despoiling landscapes, granting incomparable value status to unique mountainscapes and promoting sustainable tourism.

Coastal resources are both natural and man-made and include coral reefs, mangroves, beaches, coastal infrastructure and heritage sites. They provide homes for fisherfolk and are sites with great potential for tourism. Steps towards conservation of coastal resources include sustainable management of mangroves, regeneration of coral reefs, addressing issues of sea rise and vulnerability to climate changes, and adopting an integral approach to Integrated Coastal Management by addressing connections between coastal areas, wetlands and rivers in relevant policies and programs.

## 10. Pollution Abatement

Pollution is the inevitable generation of waste streams from the production or consumption of anything which directly impacts the quality of the medium it exists in, like air, water or soil, and indirectly impacts receptors of such a medium, like living beings. It is generally better and cheaper to prevent pollution than try to treat it after it takes place. Moreover, as certain sections of society may be more vulnerable to the impact of pollution even though they are not the generators, it has important implications for social justice.

### 10.1 Air Pollution

Air pollution is caused by use of fossil energy and industrial processes, consumption activities, inefficient pricing of fossil based use of energy, and indoor air pollution due to inefficient biomass stoves. Steps taken to abate air pollution would be an integrated approach to energy conservation and adoption of renewable energy technologies, strengthening monitoring emission standards, actions plans to address air pollution in cities through incentives and other measures, strengthening substitution of fossil fuels by bio fuels as much as possible, accelerated dissemination of improved fuelwood stoves and solar cookers suited to local cooking practices and biomass resources, and promoting research and development activities towards new technologies in harmony with nature.

### 10.2 Water Pollution

Pollution of water in surface water bodies such as rivers and wetlands, groundwater and coastal areas is to be mitigated by implementing private and public partnership models for setting up and operating effluent and sewage treatment plants, action plans and regulatory systems to address water pollution in major cities, measures to prevent pollution of water bodies from other sources, especially waste disposal, promoting R&D in development of low cost technologies for sewage treatment, taking account of groundwater pollution, dissemination of agronomy practices and encouraging Integrated Pest Management (IPM) and use of biodegradable pesticides.

### 10.3 Soil Pollution

Management of industrial and municipal waste is the major cause of soil pollution, which is to be addressed by developing and implementing strategies for cleaning up toxic and hazardous waste dumping, especially in industrial areas and abandoned mines and reclaiming such sites for future, sustainable use. Steps must be taken to empower and support local bodies to develop and implement efficient segregation, recycling and reuse of waste. Promotion of organic farming and encouraging use of biodegradable and recyclable substitutes for non-biodegradable substances are other effective measures.

## 10.4 Noise Pollution

It has now been well established that persistent exposure to high noise levels adversely affects health. To control noise pollution, noise standards must be set as appropriate to the environment. For instance, the acceptable noise level in rural areas would differ from that in urban areas. Dialogue between local bodies and religious and community representatives must be encouraged to ensure regulation of noise from use of loudspeakers and fireworks during celebrations and public functions.

## 10.5 Conservation of Man-made Heritage

Preservation of man-made heritage sites are outside the scope of the NEP but impact on environmental quality is an environmental concern. Heritage sites represent culture and are entities of incomparable value that often provide sources of livelihood to many because of their tourism potential. In order to protect them from pollution or inundation due to development projects, stricter environment standards must be set for them than those for other places. Integrated regional development plans should be made and local communities involved in ensuring that polluting activities are eliminated or minimized to maintain the heritage ambience of such sites.

## 10.6 Climate Change

Emission and increasing concentration of Greenhouse Gases in the atmosphere due to fossil fuel use, industrial and agricultural activities and deforestation, threaten to cause major upheavals in ecosystems and global climate. Developing countries like India are more vulnerable to these adverse impacts even though they are less responsible for the increase in atmospheric GHG concentration. Global climate changes are to be addressed by resorting to multilateral approaches and equal per capita entitlement of global resources to all countries.

Action plans for coping with climate changes within the country involve identifying areas of vulnerability and assessing the impact of climate change on water bodies, water resources, forests, coastal areas, agriculture, and health. The need for adaptation to future climate change must be incorporated in relevant programs for watershed management, coastal zone planning and regulation, forestry management, agricultural technologies and practices, and health programs. The Indian industry must be encouraged to participate in the Clean Development Mechanism (CDM) and enter into voluntary partnerships with other countries to address issues of sustainable development and climate change, consistent with the provisions of the UN Framework Convention on Climate Change.

## 11. Environmental Standards

Environmental standards include "ambient standards", the acceptable levels of environmental quality parameters at different locations and "emission standards" that apply to permissible levels of waste discharge in different kinds of activities. These standards are not universal but set in accordance with the specific context or situation to which they apply. They are also flexible and dynamic. For instance, the standards of a developing nation are bound to change over time.

Effective environmental standards are ensured by setting up mechanisms for regular reviews and updating with current scientific knowledge and changing circumstances and ensuring cooperation between the impacted communities and industry associations.

## 12. EMS, Ecolabelling and Certification

Monitoring and enforcement of prescribed emission standards are to be realized by Environmental Management Systems (EMS), such as ISO 14000, that demand and certify the adoption of standardized environmental management practices. Further steps would be towards ecolabeling and certification would be to promote industry associations so as to encourage the espousal of ISO 14000 amongst their members, encourage the adoption of EMS by procurement preference for ISO 14000 products, devise "Good Practice Guidelines" for ecolabels and encourage the common credit of Indian and global ecolabels, sticking to the Good Practice Guidelines.

## 13. Clean Technologies and Innovation

Clean technologies minimize generation of waste in production and recycle waste from other production process instead of treating the waste after generation as in conventional technologies. Clean technologies use less of raw materials and energy than conventional technologies and are therefore, cheaper. The challenges faced in switching over to clean technologies are that they are patented abroad, and they require setting up of comprehensive R&D towards commercially viable clean technologies. These challenges are to be met by empowering the clean technology switchover project proposals to promote capacity building in the financial sector, establishing mechanisms for networking between private and public research institutions for cooperation in technology research and development of clean technologies, and promoting adoption of clean technologies by the industry through regulatory and fiscal measures and standards setting.

## 14. Environmental Awareness

Awareness regarding environmental issues is to be raised through environmental education and ensuring access to environment information. Education about environment can be a combination of formal and informal and would be achieved through focused groups and the general public. Environmental education can be taught as a subject in schools and colleges as the Supreme Court has also ruled that environmental education must be imparted at all levels including the formal education system.

## 15. Partnerships and Stakeholder Involvement

Various agencies will be encouraged to form partnerships to achieve the policy objectives through possible partnerships such as partnerships between public institutions, local communities, private providers and voluntary agencies can take care of public functions such as monitoring environmental quality or afforestation.

## 16. Capacity Building

Ongoing building of capacity of all institutions concerned with achieving the policy objectives will be realized by reviewing the present institutional capacities of bodies at all levels of government, and preparing and implementing programs for enhancing their capacities. Funds must be earmarked for capacity building programs and continuous upgradation of knowledge and skills of all the personnel involved in environment management.

## 17. Research & Development

A scientific understanding of environmental issues can only be achieved through focused research by academic institutions, scientists, researchers, government and private institutions in the key areas of taxonomies of living natural resources, ecological processes and pathways, technologies for environmental management and clean production. This will be facilitated by periodically identifying research areas and establishing research programs to be conducted within and outside the government with financial and institutional support.

## 18. International Cooperation

Since 1972, India has participated in various multilateral agreements with other countries on key global environmental management and environmental cooperation and complied with its commitments. To enhance its capacity to comply with its commitments, it will participate in multilateral programs for enhancing flow of resources for sustainable development and provide assistance to other countries for environment management.

The objectives, principles, strategies and actions listed in the policy are dynamic and meant to evolve through regular reviews. Moreover, to ensure effective implementation of the policy, mechanisms will have to be put in place to carry out periodic reviews for each of the action plans. Besides this, a cabinet or a nomination of the cabinet will conduct an annual review of the policy and its implementation to ensure accountability of all those involved in the process. These reviews will also address practical issues and the findings will be publicly disclosed.

The National Environment Policy is a testament to the exhaustive process that has gone into its formulation. The effort and inputs of various key individuals and institutions from the field of research, finance, technology and environmental bodies have gone into making the policy an effective guide to action towards environmental conservation and sustainable development.

*(Dr. Usha K R is a Language Editor at the Icfai Business School Research Center, Chennai. She can be reached at ushakr@gmail.com)*

# 6

# Environmental Governance in India

*Geetanjoy Sahu*

***Over the years, there has been an increasing awareness and consciousness among people about their right to a healthy environment. Like any other socioeconomic and political problem, environmental problem has caught the attention of policymakers, intellectuals, social activists and research scholars. No nation can afford to ignore the emerging environmental problems such as depletion of the ozone layer, acid rain, green house effect, soil erosion, deforestation, water pollution, air pollution, noise pollution, etc. To identify and solve these problems, a number of actors such as international and national institutions, civil society, environmental groups, media and local people are involved. This involvement has led to the development of the concept of environmental governance. Environmental governance deals with the manner in which decisions concerning the environment are taken, and the participants in the decision-making process. This paper studies the evolution of the concept of environmental governance in India. In doing so, it highlights the contributing factors and emerging issues in the environmental governance process. It also explores the significance and implications of these issues to ensure a healthy environment and aims to protect and improve the environment.***

*Source: The Icfai Journal of Environmental Law, July 2006.* 

## Introduction

In the last several decades, there has been an increasing concern for environmental problems throughout the world involving depletion of ozone layer, acid rain, green house effect, soil erosion, deforestation, water pollution, air pollution, noise pollution, etc. The frequency of these problems is largely related to the nation's quest for development, technological advancement, industrialization and urbanization which place unprecedented demands on the regenerative capacity of the ecosystems and jeopardize the conservation of environment. These environmental problems have had devastating impact on human well-being and are also raising the specter of irreversible long-term damage to the ecosystem. This has led to the emergence of various movements and campaigns to protect the environment and to ensure sustainable use of resources. As in other parts of the world, the concern for environmental protection and regulation in India has emerged in the seventies and assumed public appeal in the subsequent years. Though the genesis of concern for environmental protection in India can be traced back to the early 20th century when people protested against the commercialization of forest resources during the British colonial period, it was only in the 1970s that it received public attention (Guha, 1991).

The UN Conference on the Human Environment was held at Stockholm in June 1972, which focused on the evolving principles and action plans for controlling and regulating environmental degradation. Since this conference, there have been many conferences and agreements not only at the international level but also at the state level directed towards the protection of environment and halting of environmental degradation. India as a signatory to many of the international environmental agreements has enacted a number of environmental laws and policies to protect and improve its environment.

In the same manner, realizing better environmental decision-making requires cooperation at all levels and between all stakeholders, the federal government's 1992 policy statement on environment and development acknowledged the role of various informal actors like environmental NGOs/groups, civil society, people's participation, media, etc., in the environmental decision-making process. The involvement of various actors in the environmental decision-making process has led to the development of the concept "environmental governance". It concerns interactions among formal and informal institutions within society that influence how environmental problems are identified, framed and dealt with. The current state of environmental governance in India can be understood by a close examination of the following aspects:

## Legal Framework for Environmental Protection

Legislative efforts to protect the environment in India date back to the mid-nineteenth century. Many of these Acts dealt with environmental regulation in a piecemeal manner and proved ineffective in reducing the levels of pollution. Action against polluters had necessarily to be initiated in courts by those affected. Pollution and environmental degradation were addressed very generally in terms of nuisance, negligence, liability, and a few principles of tort law (Thakur, 1997). The pre-Independence laws had not dealt with environmental protection exclusively. For example, the Indian Penal Code (IPC), 1860, had a chapter (chapter XIV) which dealt with offences affecting public health, safety and convenience, and covered aspects like water, air and noise pollution, whereas the post-Independence laws, deal exclusively with environmental protection (Mehta, 1991).

In the post-Independence India, many positive trends can be observed in environmental governance. Environmental laws were strengthened, particularly in the 1970s and again in the 1990s. In India, the need to integrate environmental concerns into the process of economic development was voiced as far back as the late 1960s, during the formulation of the Fourth Five-Year Plan (1969-74), which stated that "planning for harmonious development is possible only on the basis of a comprehensive appraisal of environmental problems". Integration of environmental resource management with national economic planning started with the Sixth Five-Year Plan (1980-85).

Similarly, inspired by the Stockholm Declaration of 1972, the Water (Prevention and Control of Pollution) Act, 1974 (the Water Act), provided for the institutionalization of pollution control machinery by establishing Boards for prevention and control of pollution of water. These Boards were entitled to initiate proceedings against infringement of environmental law, without waiting for the affected people to launch legal action. The Water Cess Act, 1977, supplemented the Water Act by requiring specified industries to pay cess on their water consumption. With the passing of the Air (Prevention and Control of Pollution) Act, 1981 (the Air Act), the need was felt for an integrated approach to pollution control. The Water Pollution Control Boards were authorized to deal with air pollution as well, and became the Central Pollution Control Board (CPCB) and the State Pollution Control Boards (SPCBs) (Divan and Rosencranz, 2001).

The Bhopal Gas leak disaster of December 1984 precipitated the tightening of environmental regulation. In 1985, the Department of Environment was changed into Ministry of Environment and Forests (MoEF) and given greater powers (Dwivedi, 1997). The Environment (Protection) Act, 1986 (EPA), was passed, to act as an umbrella legislation. The Act also vested powers with the central government to take all measures to control pollution and protect the environment. The Environment (Protection) Rules, 1986, were subsequently notified to facilitate the exercise of powers conferred on the Boards by the Act. The EPA identifies the MoEF as the apex policy making body in the field of environment protection. The MoEF acts through the CPCB and the SPCBs. The CPCB is a statutory organization and the nodal agency for pollution control. The EPA in 1986 and the amendments to the Air and Water Acts in 1987 and 1988, furthered the ambit of the Boards' functions. This umbrella law empowers the central government to decide emission and effluent standards, restrict industrial sites, promulgate procedures and safeguards for accident prevention and handling of hazardous waste, investigate and research pollution issues, conduct on-site inspections, establish laboratories, and collect and disseminate information. The Seventh and Eighth Five-Year Plans recognized the issues of environmental resource preservation and sustainability as being as important as many other developmental objectives. The policies enunciated in the National Conservation Strategy and Policy Statement on Environment and Development and the Policy Statement on Control of Pollution, both established in 1992, were pursued in the Ninth Five-Year Plan (Vyas and Reddy, 1998). Assigning conservation a high priority, both at the Central and State levels, the Tenth Five-Year Plan (2002-07) also seeks to tackle the environmental degradation in a holistic manner in order to ensure both economic and environmental sustainability.

Similarly, there has been a considerable growth in the field of environmental laws in the 1990s. The enactment of various laws to protect and improve the environment, namely, the Public Liability Insurance Act, 1991, the Wild Life (Protection) Rules, 1995, the Wild Life (Specified Plants-Conditions for Possession by Licensee) Rules, 1995, the National Environment Tribunal Act, 1995, the National Environment Appellate Authority Act, 1997, the Biodiversity Act, 2002, etc. have further strengthened and expanded the environmental governance process by addressing different natures of environmental problems and seeking the incorporation of various groups in the environmental decision-making process. In addition to these, a number of State Governments have enacted special laws for protecting and improving the environment.

## Constitutional Provisions

Environmental protection has found a special mention in the Indian Constitution. In fact, the environment protection has been given a constitutional status in the Indian polity. The fundamental rights and the directive principles of state policy underline India's commitment to protect and improve the environment. In terms of the directive principle of state policy, by the 42nd Amendment of 1976, two new Articles were inserted: Article 48-A and Article 51-A (g). The former, under the Directive Principles of State Policy, makes it the responsibility of the State Government to protect and improve the environment and to safeguard the forests and wildlife of the country. The latter, under Fundamental Duties, makes it the fundamental duty of every citizen to protect and improve the natural environment including forests, lakes, rivers and wildlife and to have compassion for living creatures.

Similarly, Part III of the Constitution of India incorporates fundamental rights which have been made judicially enforceable. The Supreme Court of India has contributed significantly especially since 1980s onwards in broadening the contents and contours of some of these basic rights. In this connection, it is worthwhile to refer to the decision of the Apex Court in Dehradun Quarry's Case, where the Supreme Court moving under Article 32 ordered the closure of some of the quarries on the ground that these were upsetting the ecological balance and held that environmental degradation violates the fundamental right to life under Article 21. (A detailed discussion on the role of Supreme Court in environmental governance has been made in the later part of this article).

Therefore, the Constitution has made double provisions, which include directing the State for the protection and improvement of environment and expecting citizens to help in the preservation of natural environment. These provisions are indicative of government's awareness of a contemporary problem and of the need for providing a constitutional base for further action at the national, state and local levels.

## Integration between Environmental Policy and Economic Planning

Since the release of the Brundtland Commission's report, Our Common Future, the concept of sustainable development has captured the world's attention and emerged as the new political ideology to be addressed. The emergence of the concept 'sustainable development' has fundamentally changed the nature and scope of the debate on the environment and its relation to development. Before the evolution of this concept, whenever a new situation emerged, it was added to the concept of economic development. Sustainable development, on the other hand, has actually subsumed the entire notion of economic development within its orbit. Now, the pursuit of economic growth is no longer considered the core value; rather it is a part of the larger picture, the central theme of which is how to integrate economic and environmental concerns in a development strategy (Dwivedi, 1997).

As far as integration between environmental policy and economic planning in India is concerned, environment was not given high priority and sufficient recognition in the plans until 1968 (Kaur, 1992). After independence, in the process of nation-building, India continued the pursuit of economic development based on modern industry and agriculture with comprehensive centralized planning. Believing that industrialization would promote economic interdependence, and that, in turn, would tie the country together as a nation, the then government adopted a central planning approach to distribute economic growth among the regions.

However, since 1970s, there has been an increasing concern for environmental protection and improvement in Indian policy and planning process. It was in the Fourth Five-Year Plan (1969-74) that for the first time, environmental issues were given high priority and were highlighted in the following words:

"It was an obligation for each generation to maintain the productive capacity of land, air, water and wild-life in a manner which leaves its successors some choice in the creation of a healthy environment. The physical environment is a dynamic, complex and inter-connected system in which any action in one part affect the others. There is also interdependence of human beings and nature, and harmonious development recognizes this unity of nature and man. Such planning is possible only on the basis of a comprehensive appraisal of environmental issues, particularly, economic and ecological. There are instances in which timely, specialized advice on environmental aspects could have helped in project design and in averting subsequent adverse affects on the environment, leading to loss of invested resources. It is necessary, therefore, to introduce the environmental aspect into our planning and development. Along with effective conservation and rational use of natural resources, protection and improvement of human environment is vital for national well-being."

In the subsequent five-year plans, efforts have been made to ensure that the pursuit of development goals does not lead to reduction in the quality of life through deterioration in environmental conditions. It is, therefore, necessary that in the rational development of industrial activity and in the utilization of natural resources, due weight is given to the impact of such activity on the environment. Hence, it appears that Indian planners have come around to realize that development cannot subsist upon a deteriorating environmental resource base and that environment cannot be protected when growth leaves out of account the costs of environmental destruction.

The laws and policies relating to the protection and improvement of environment reflect the growing concern among the policymakers to tackle the emerging environmental problems. Since the 1972 UN Conference on Human Environment, considerable attention has been accorded to policy, legislative and economic instruments in India to ensure that the periodic planning and development policy do not threaten the overall move towards environmental sustainable development. But, the biggest problem that India is facing is the administration and implementation of the various programs. While the country has adequate legal mandates to solve environmental problems, the gaps in policy implementation mechanism indicate that the enforcement policy is rather weak, and at times non-existent. Despite a multitude of legislation, constitutional directives and duties and the setting up of Pollution Control Board all over the country, the success in curbing environmental degradation has not been very encouraging because of failure in implementing environmental laws. Meanwhile, the judiciary, adorning the mantle of an ombudsman, has filled the lacunae in the laws and enforced their implementation, by its landmark and unconventional decisions.

## Role of Judiciary in Environmental Governance

Since the last quarter of the 20th century, the Indian Judiciary has been playing a significant role in the environmental governance process. While conventionally, the executive and the legislature play the major role in governance, the Indian experience, particularly in the context of environmental governance, is that the judiciary has begun to play a vital role in protecting the environment and checking its degradation and pollution. Perhaps no judiciary in the world has devoted as much time, effort and innovativeness in protecting environment as the judiciary of India has for the last two decades. In the process, it has reinterpreted some of the Environmental Acts, created new institutions and structures and conferred additional powers on the existing ones.

The intervention of judiciary in environmental governance process is necessitated due to ineffectiveness in the enforcement of existing environmental laws by the implementing agencies. As indicated above, India has employed a range of regulatory instruments to preserve and protect its environment. There are over 200 Central and State statutes, which have at least some concern with environmental protection, either directly or indirectly (Divan and Rosencranz, 2001). But, failure on the part of the government agencies to effectively enforce environmental laws and non-compliance with statutory norms by polluters has resulted in accelerated degradation of environment. Most of the rivers and water bodies are polluted and large-scale deforestation is being carried out with impunity. There has also been a rapid increase in casualties due to respiratory disorders caused by widespread air pollution. Such large-scale environmental degradation and adverse effects on public health prompted the environmentalists as well as non-governmental organizations to approach the courts, particularly the higher judiciary, for suitable remedies (Ramesh, 2002). To draw the attention of the judiciary to environmental cases, public-spirited persons have used Public Interest

Litigation (PIL) as one of the important instruments ever since it emerged in the Indian legal system.

Prior to the emergence of the concept PIL, Criminal Law provisions as contained in the Indian Penal Code, Civil Law remedies under the law of torts and provisions of the Criminal Procedure Code had been providing effective and preventive remedies for public nuisance cases including air, water and noise pollution (Thakur, 1997). However, due to lack of people's awareness about the environmental problems and limited knowledge of environmental laws, there were problems in drawing the attention of the judiciary towards environmental problems. Again, there was no provision for allowing the third party to seek the help of the judiciary if the party was not directly affected by environmental problems. With the emergence of PIL, now, any citizen of India can file a petition on behalf of any citizen or group of citizens who themselves are incapable of approaching the court.

As a result, we find a number of litigations on environmental issues. Right from the Dehradoon lime stone quarrying case to the *T N Godavarman Thirumulkpad vs. Union* of India case, there are a plethora of environmental cases like the Ganga Water Pollution, Delhi Vehicular Pollution, Oleum Gas Leak, Narmada Dam Case, Tehri Dam Case, Vellore Industrial Pollution, Patancheru Industrial Pollution, etc., which have come to the judiciary through PIL. These petitions have been filed either by individuals, voluntary organizations, or by letters sent to the judges. The overview of some of the cases will make us understand the contribution of judiciary in Indian environmental policymaking process.

In the Dehradun Quarrying Case, which was the first case of its kind in India where judiciary intervention in environmental cases involving issues relating to environment and ecological balance was seen. It related to closing down of large number of leases of lime-stone quarries in Mussoorie-Deheradun region which were polluting the environment, causing ecological imbalance and hazard to the health of not only human beings but also of all inanimate and animate things. This case brings into sharp focus the conflict between development and conservation, and serves to emphasize the need for reconciling the two in the larger interest of the country. In this case, for the first time, the Supreme Court held that the fundamental right to a wholesome environment is a part of the fundamental right to life in Article 21 of the Constitution. The Court in its judgement held that all fresh quarrying should be stopped on the ground that their operations were upsetting the ecological balance. This view is also supported by Justice Singh's concluding observations justifying the closure of polluting tanneries in the Ganga Pollution case: To quote the judgement from the bench, "We are conscious that closure of tanneries may bring unemployment, loss of revenue, but life, health and ecology have greater importance to the people."

Many High Courts have also explicitly recognized right to a wholesome environment as a dimension of right to life guaranteed in Article 21. In this context, the observations of Andhra Pradesh High Court in *T Damodhar Rao vs. SO Municipal Corporation,* Hyderabad (Leelakrishna, 1987) while considering a writ petition to enjoin the Life Insurance Corporation and the Income Tax Department from building residential houses in a recreational zone, are worth noting here:

"It would be reasonable to hold that the enjoyment of life and its attainment and fulfilment guaranteed by Article 21 of the Constitution embraces the protection and preservation of nature's gifts without which life cannot be enjoyed. There can be no reason why practice of violent extinguishments of life alone should be regarded as violation of Article 21 of the Constitution. The slow poisoning by the polluted atmosphere caused by environmental pollution and spoilation should also be regarded as amounting to violation of Article 21 of the Constitution."

It has become the legitimate duty of the Courts as the enforcing organs of constitutional objectives to forbid all actions of the State and the citizen from upsetting the environmental balance. In this case, the very purpose of preparing and publishing the development plan is to maintain such an environmental balance. The objective of reserving certain areas as a recreational zone would be utterly defeated if private owners of the land in that area are permitted to build residential houses. It must, therefore, be held that the attempt of the Life Insurance Corporation of India and the Income Tax Department to build houses in this area is contrary to law and also contrary to Article 21 of the Constitution.

The fact that the courts are aware of the weakness in our environmental legislation is evident from several other landmark judgments in which they took upon themselves the task of expanding the ambit of Article 21 of the constitution by interpreting that right to life included the right to live in a healthy environment. The Court has proclaimed that environment protection is a constitutional mandate and courts cannot sit with their eyes closed when assaults on environment caused by adverse socioeconomic policies pose a threat to ecology.

From the aforesaid cases of judiciary intervention in environmental issues, it is observed that the Court has not only reminded the executive and administration their Constitutional duties to protect and preserve the environment but also evolved new principles and policies by way of expanding right to life and balancing between right to environment and right to development. This has led to fundamental changes in the pattern of environmental governance in India.

## Informal Actors in Environmental Governance

Although there is diversity in governmental structures, environmental governance has not fully developed in India because the Central Government remains the most important actor in environmental governance process of the country. It still plays a strong role in Indian environmental governance for two reasons. First, the nature of environmental pollution itself makes public intervention, to some extent, indispensable for dealing with environmental problems. Thus, market mechanisms are not always a panacea for solving environmental problems. Second, these environmental governance structures were established at a time when India adopted centralized political regimes in which the Central Government held a dominant position over local governments. The power of local governments was limited, and in many cases, the Central Government had the power to designate local governors. After Independence, centralization of power might have been an unavoidable choice for India in the quest of nation-building process and achieving the developmental goals.

However, with the emergence of societal forces in the early 1970s, the state-centric development policy faced formidable challenges. Among the societal forces, environment movements challenged the ideology of economic development which had not taken environmental factors into consideration and advocated for the sustainable use of natural resources. The ideology of economic development, which remained almost monolithic in the first two decades of policymaking process, faced a major foundational challenge from the environmental groups, which consciously emphasized sustainable use of natural resources (Shiva, 1991). With the emergence of environmental movements, there is a change in the political and economic attitudes and practices in several important ways. Modifications to social cost-benefit analysis, the onset of environmental impact assessment and environment auditing, risk analysis, public inquiries, new legislative measures, plus the successful political-legal activities of non-governmental groups have all helped to give policies and actions an environmental color (Sinha, 1998).

India has witnessed quite a wide range of environmental movements such as *Silent Valley Movement* in Kerala, *Chipko Movement* in Uttar Pradesh, *Aapiko Movement* in Karnataka, *Narmada Bachao Andolan* in Central India, *Gandhamardhan Movement* in Orissa, etc. These movements aim at halting environmental degradation or bringing about environmental restoration or regeneration or sustainable use of natural resources.

In addition to the above actors, many new environmental actors emerged and contributed to the progress of Indian environmental governance in the 1990s. A good example is the Civil Society Movement in India. This movement reflecting the people's dissatisfaction with the government's development planning as well as with its unsustainable model of development, has shifted the debate development. Civil society, through bottom-up advocacy, has put environment on the development agenda. The Federal Government's 1992 policy statement on environment and development acknowledged the role of various informal actors like environmental NGOs/groups, civil society, people's participation, media, etc. in the environmental decision-making process. As a result, numerous NGOs have come into existence in India, and their activities have covered various aspects of environmental problems.

Apart from the environmental groups/NGOs, the media has been playing a significant role in environmental governance process of India by addressing various environmental problems and its negative impact on the health of the people and on the eco-system. For example, in the Delhi Vehicular Pollution Case, the media had highlighted various impact of air pollution on human health and drew the attention of the judiciary to protect the environment and the health of the people.

## Conclusion

As in other parts of the world, concern for environmental protection has been increasing in India since the last two decades. There has been significant changes in the policy making process to tackle environmental degradation and attempts have been made to consider environment in the formation of development plans and programs. Moreover, new actors such as environmental NGOs and the media have also been emerging, particularly during the 1990s. The interactions between formal and informal actors in designing and formulating policies have expanded the ambit and contours of environmental governance process in India. In contrast to the past dominance of formal institutions, a combination of both formal and informal campaigns for environmental protection has to some extent succeeded. New types of environmental programs, which involve various actors other than the Central Government, also have been initiated in India.

In the same manner, the Indian Judiciary has been playing a significant role in the environmental governance process of India. The intervention of judiciary has ensured right to a healthy environment by expanding the meaning of the fundamental right to life, which includes right to a healthy environment, livelihood and health under Article 21 of the Indian Constitution. This has led to fundamental changes in the pattern of environmental governance in India.

Similarly, the potent influence of initiatives undertaken by international agencies is broadly felt in the environmental policy processes of India. India as a signatory to many international agreements on environmental protection has introduced policy mechanisms to control pollution and halt degradation of environment.

To conclude, it can be said that environmental governance process in India has expanded the meaning and nature of environmental decision-making process in terms of issues being raised as well as the involvement of actors in the process. Again, institutional style similar to those of developed countries are increasingly being adopted in India, and this may become the predominant trend. Though environmental governance and policies of India are likely to become more similar in style to those in industrialized countries, the actors and processes involved in decision-making and the implementation of environmental governance are not always the same as in developed countries. This suggests that more in-depth analysis of environmental governance system in India is necessary.

*(Geetanjoy Sahu is a Research Scholar, Development Administration Unit, The Institute for Social and Economic Change (ISEC), Bangalore, India. The author can be reached at geetanjoy@rediffmail.com)*

## References

1. Divan, S and Rosencranz A (2001), *Environmental Law and Policy in India*, New Delhi: Oxford University Press.
2. Dwivedi, O P (1997), *India's Environmental Policies, Programs and Stewardship*, London: Macmillan Press Ltd.
3. Guha, Ramchandra (1991), *Environmentalism: A Global History*, New York: Oxford University Press.
4. Kaur, Dilbar (1992), "Environmental Protection in India: Constitutional Conspectus", in Paras Diwan and Peeyushi Diwan (eds.) *Environment Administration, Law and Judicial Attitude*, New Delhi: Deep & Deep Publications.
5. Kohli, Atul (1994), *State Power and Social Forces: Domination and Transformation in the Third World*, Cambridge: Cambridge University Press.
6. Leelakrishna, P (1987), "Law and Policy Relating to Environment", in Paras Diwan and Peeyushi Diwan (eds.) *Environment Administration, Law and Judicial Attitude*, New Delhi: Deep & Deep Publications.
7. Mehta, Chetan (1991), *Environmental Protection and the Law*, New Delhi: Ashish Publishers.
8. Ramesh, M K (2002), "Environmental Justice: Courts and Beyond", *Indian Journal of Environmental Law*, June, Vol. 3, No. 1.
9. Shiva, V (1991), *Ecology and Politics of Survival: Conflicts over Natural Resources in India*, New Delhi: Sage Publications.
10. Sinha, P C (1998), *Green Movements*, New Delhi: Anmol Publications Pvt. Ltd.
11. Thakur, Kailash (1997), *Environmental Protection Law & Policy in India*, New Delhi: Deep & Deep Publications.
12. Vyas, V S and Ratna Reddy V (1998), "Assessment of Environmental Policies and Policy Implementation in India", *Economic and Political Weekly*, January 10.
13. World Commission on Environment and Development (WCED) (1987), *Our Common Future*, New York: Oxford University Press.

## Reports and Documents

1. Planning Commission, The Government of India, Fourth Five-Year Plan, 1969-74.
2. Planning Commission, The Government of India, Sixth Five-Year Plan, 1980-85.
3. Planning Commission, The Government of India, Ninth Five-Year Plan, 1997-2002.
4. Planning Commission, The Government of India, Tenth Five-Year Plan, 2002-07.
5. *Rural Litigation and Entitlement Kendra vs. State of Uttar Pradesh,* AIR 1987 SC 395.

# 7

# A Public Health Approach to Environmental Protection in India

*Krishnayan Sen*

***A public health approach to environmental protection potentially touches upon all spheres of human activity, and claims to override or trump other considerations. The UN Conference on Human Environment declared that 'Man has the fundamental right to freedom, equality and adequate conditions of life in an environment of a quality that permits a life of dignity and well-being'. Article 6 of the International Covenant on Civil and Political Rights, 1966, provides that everyone has the right to life. The Human Rights Committee has consistently sought information on such measures in the public health and environmental fields, including the registration of transportation and dumping of nuclear waste. A public health approach has led the Supreme Court to derive, adopt and apply a range of principles and standards to guide the development of environmental jurisprudence.***

*"Whatever I dig from thee, O Earth, May that have quick growth again. O Purifier, May we not injure thy vitals or thy heart."*

– Atharva Veda 12.1.35

*Source: The Icfai Journal of Environmental Law, October 2004.* 

## I. Introduction

The late 20th century has witnessed an unprecedented increase in legal claims and awareness for both public health rights and environmental standards. Seldom have legal remedies stood so squarely in the centre of wider social movements for human and environmental protection. In recent years, the law and policy-making in the environment and public health sectors, at both international and domestic levels, have been marked not only by speed and proliferation, but also by remarkable innovation. A public health approach to environmental protection potentially touches upon all spheres of human activity, and claims to override or trump other considerations.

The first part of the paper addresses environmental rights affecting public health endeavor in the international context. After analyzing the international treaties and conventions relating to environmental and public health protection, the author would seek to locate the right to health as a norm in customary international law. The second half of the paper examines emerging trends in the application of environmental rights in domestic law. This part would enunciate the position of the right to health and environment under the Indian Constitution and submit as to how the enjoyment of one is intrinsically linked with that of the other. It would be observed that while much of the current legal debate occurs at the international level, the detailed experience of environmental rights in domestic jurisdictions (India being the case in point) offers rich instruction, simultaneously showing promising ways to move forward as well as examples of mistakes to be avoided. The equal emphasis on international and domestic law is, therefore, deliberate. The paper would conclude with the author's final submissions and recommendations.

## II. Role of International Human Rights Law in a Public Health Approach to Environmental Protection

What role should international human rights law play in the protection of the environment? At Stockholm, in 1972, the UN Conference on Human Environment declared that *'Man has the fundamental right to freedom, equality and adequate conditions of life in an environment of a quality that permits a life of dignity and well-being'.*[1] Twenty years later, at the Rio Conference on Environment and Development 1992, this initial emphasis on a human rights approach, more importantly on a public health perspective, has not been maintained. Avoiding the terminology of rights altogether, the Rio Declaration merely asserts that *'Human beings are the center of concerns for sustainable development. They are entitled to a healthy and productive life in harmony with nature'.*[2]

The Rio Declaration's failure to give greater explicit emphasis to public health rights is indicative of continuing uncertainty and debate about the proper place of public health law in the development of international environmental law.[3] This is not because of any lack of interest in the topic. On the contrary, references to a right to a decent, healthy, or viable environment have appeared in several global or regional human rights treaties,[4] and in declarations or resolutions of international organizations.[5]

Some effort has also been made by human rights institutions[6] and by noted scholars,[7] to derive environmental rights from other internationally protected rights, such as life or property. There is, moreover, a growing trend to give environmental protection constitutional status in many national legal systems, either explicitly or by judicial interpretation of other constitutional guarantees.[8] These developments have encouraged the UN Sub-Commission on the Prevention of Discrimination and Protection of Minorities to undertake a study of human rights, notably public health rights and the environment.[9]

The Final Report[10] of the Sub-Commission in 1994 (hereafter, the *'Ksentini Report')* offers a conception of human rights and the environment, which is much closer to Principle 1 of the 1972 Stockholm Declaration[11] than to Principle 1 of the 1992 Rio Declaration.[12] Its conclusions are based on a survey of national and international human rights law and of international environmental law.[13] The report's most fundamental conclusion is that there has been 'a *shift from environmental law to the right to a healthy and decent environment*'. This right, it argues, is part of existing international law and is capable of immediate implementation by existing human rights bodies. Its substantive elements include the right to development, life, and health, and access to effective national remedies.

It is submitted that environmental rights, by finding a place in the existing international human rights law regime, as proposed in the Ksentini Report[14], could serve three possible functions:

### (i) Participatory Rights

First, drawing on existing civil and political rights, they might be used to give individuals, groups and non-Governmental organizations access to information, remedies and political processes. This approach rests on the view that environmental protection and sustainable development cannot be left to governments alone but require and benefit from notions of civic participation[15] in public affairs already reflected in existing civil and political rights.[16] Principle 10 of the Rio Declaration[17] does give substantial support in mandatory language for participatory rights of a comprehensive kind. It provides:

"Environmental issues are best handled with the participation of all concerned citizens, at the relevant level. At the national level, each individual shall have appropriate access to information concerning the environment that is held by public authorities, including information on hazardous material and activities in their communities, and the opportunity to participate in the decision-making processes...Effective access to judicial and administrative proceedings, including redress and remedy, shall be provided.[18]

The real test of Principle 10's significance lies in international instruments rather than in national law. It is here that the most important applications of the principle have taken place, most notably in India after the *Bhopal disaster.*[19]

### (ii) Environmental Quality

A second possibility is to treat a decent, healthy or viable environment as an economic or social right, comparable to those already protected in the 1996 ESCR Covenant.[20] It is submitted that this approach would give environmental quality a comparable status to other economic and social rights, with some priority over non-rights-based objectives, and that it would recognize the vital character of the environment as a basic condition of life, indispensable to the promotion of public health and welfare.[21]

### (iii) The Environment as a Solidarity Right

The third option of treating environmental quality as a solidarity right would mainly entail other governments and organizations in cooperating to provide the necessary resources, skills and technology to achieve the realization of environmental objectives.[22] Its main beneficiaries would be developing countries, like India, whose participation in environmental treaties is particularly desirable if global coverage is the objective, but it would scarcely be a 'human' right in any orthodox sense. The expression of the concept *"common but differentiated responsibility"*[23] lies at the heart of both Rio Conventions and the Montreal Protocol to the Ozone Convention 1987,[24] and for identical reasons: Developed states have a strong interest in persuading developing states to participate, developing states have much less of an interest in doing so unless there are clear benefits for them.[25] It is, therefore, submitted that although not put in terms of solidarity rights, the notion of solidarity is clearly central to this part of international environmental law, and is recognized as such in the interdependence of commitments made on both sides.

## III. Environmental Rights and Standards in Existing Human Rights and Public Health Treaties

The paper may now examine as to what extent environmental rights are found in existing human rights and public health treaties. It is, thus, considered only with *lex lata,* and not, as in the preceding discussions, with *lex ferenda.*[26]

### III.1 Treaties Concerned with Civil and Political Rights

*The Right to Life:* Article 6 of the International Covenant on Civil and Political Rights, 1966[27] provides that everyone has the right to life.[28] It is evident that this right involves at least a prohibition on the state not to take life intentionally or negligently. Thus, in extreme cases, the right might be invoked by individuals to obtain compensation where death resulted from some environmental disaster, like Bhopal or Chernobyl, in so far as the State is responsible. A more difficult question is whether the right goes beyond a mere prohibition not to take life and may also impose some positive obligation on the State to take steps which would prevent a reduction in, or promote, life expectancy; for example, by the provision of better drinking water or less polluted air or soil.

The Human Rights Committee has consistently sought information on such measures in the public health and environmental fields, including the registration of the transportation and dumping of nuclear waste.[29] But it is submitted that there are problems with such a broad approach, including overlap with the

International Covenant on Economic, Social and Cultural Rights; doubts as to whether these broader obligations to respect the right to life are immediate or progressive; and if the latter, problems in adjudicating on individual complaints under the Optional Protocol.[30]

Significantly though, only two cases are known to have been brought under the complaints machinery of the treaties.[31] Therefore, while the right to life appears to have some potential application to the environmental field, it has not yet been successfully invoked under the machinery for individual complaints provided by the Convention; and, indeed, does not speak of much promise for a developing country like India, where the basic awareness itself is so deplorable.

## III.2 Treaties Concerned with Economic, Social and Cultural Rights

This section will now broach the main treaties concerned with economic, social and cultural rights that embody environmental rights primarily at the global level; namely, the International Covenant on Economic, Social and Cultural Rights 1966.[32] During the course of the ensuing discussion, it must, however, be remembered that India being a party to the Covenant is subject to the following provisions and bound by international obligations.

### 111.2.1 Right to a Decent Working Environment

Article 7 of the Covenant provides that the party States *'recognizes the right of everyone to just and favorable conditions of work which ensure, in particular.. safe and healthy working conditions'*.[33] This would seem to include, for example, that an individual in the work place should be free from exposure to pollution. But to appreciate the scope of the State's obligations under Article 7, it is also necessary to consider Article 2 (1), which provides that each party to the Covenant undertake certain steps, individually and through international assistance and cooperation to the maximum of its available resources, with a view to achieving *progressively the full realization* of the rights recognised in the present Covenant by all appropriate means including, particularly, the adoption of legislative measures.[34]

### 111.2.2 Right to Decent Living Conditions

We now come to the two derivative environmental rights found in treaties relating to economic, social and cultural rights. The first of these is the right to decent living conditions. Article 11(1) of the Covenant provides that the parties *'recognize the right of everyone to an adequate standard of living.. and to the continuous improvement of living conditions. The parties States will take appropriate steps to ensure the realization of this right'.*[35] Examining the content of Article 11(1) from an environmental point of view, the obligation to improve living conditions could require States, for example, to ensure less pollution of the atmosphere and water, reduced exposure to noise pollution etc.

**III.2.3 An International Right to Health**

The second derivative right concerns health. Article 12 of the Covenant[36] provides as follows:

1. The States Parties...recognize the right of everyone to the enjoyment of the highest attainable standard of physical and mental health.
2. The steps to be taken by the States Parties...to achieve the full realization of this right shall include those necessary for...

   (a) The improvement of all aspects of environmental and industrial hygiene;

   (b) The prevention, treatment and control of epidemic, endemic, occupational and other diseases.

The realization of this right, like that to decent living conditions, may require States to reduce or eliminate various forms of pollution.[37]

The American Protocol 1988 or the San Salvador Protocol also provides in Article 10 that:

- Everyone shall have the right to health, understood to mean the enjoyment of the highest level of physical, mental and social well-being.
- In order to ensure the exercise of the right to health, the State Parties agree to recognize health as public welfare.[38]

In case of the treaties, it must be stressed that the rights that they contain are not immediate, but are to be progressively realized, the rate of progress depending on the resources of the State concerned.[39]

Health is, further, defined in the Preamble of the WHO Constitution, as a *"state of complete physical, mental and social well-being, and not merely the absence of disease or infirmity".*[40]

Even in the Universal Declaration of Human Rights, health is mentioned as an aspect of the right to an adequate standard of living.[41] Finally, mention may be made of Article 24 of the Convention on the Rights of the Child 1989[42], which provides that a child has the right to enjoy the highest attainable standard of health. Here, *for the first time in a human rights treaty, there is explicit recognition of the connection between health and the state of the environment.*[43]

## IV. Tracing the Right to Health in Customary International Law

Having examined the right to health embodied by international treaties, we must now address a further question and enquire if such a right has attained the character of a norm of customary international law. Contrary to treaty law, which is based on the consent of States and does not normally bind third parties[44], customary international law, understood as a general practice accepted as law[45], binds all States without exception and irrespective of their express consent.[46] The characterization of the right to health as customary international law would then broaden the field of application of such a right and its enforceability beyond the States parties to human rights treaties. Therefore, if it may be established that right to health is a part of the customary international law; then India has to respect the same irrespective of its being a party to a specific treaty.

As confirmed in the *Nicaragua case*[47], custom is constituted by two elements— the objective one of 'a *general practice*', and the subjective one '*accepted as law*', the so-called *opinio iuris.*[48] In the *Continental Shelf case,*[49] the court stated that the substance of customary international law must be 'looked for primarily in the actual practice and *opinio juris* of States.[50]

The answer, therefore, to the question whether or not the right to health has become customary international law depends on the analysis of the state practice in this field. We consider four criteria to be important: (a) the wide acceptance of non-binding instruments dealing with the right to health; (b) the wide participation in multilateral treaties that establish the right to health; (c) the reference to such a right in national legislation, mainly at the constitutional level; and (d) the implementation of the right to health before municipal courts. First, let us consider the non-binding instruments—the so-called *soft law*—that deal with the right to health. Their wide acceptance by the international community is to be seen as an initial evidence of the customary character of the right to health. In this context, the most relevant document to be mentioned is the Universal Declaration of Human Rights, adopted by the UNGA in 1948 with no negative votes. Article 25 of the UDHR[51] reads as follows:

- Everyone has the right to a standard of living adequate for the health and well-being of himself and of his family, including food, clothing, housing and medical care and necessary social services, and the right to security in the event of unemployment, sickness, disability, widowhood, old age or other lack of livelihood in circumstances beyond his control.
- Motherhood and childhood are entitled to special care and assistance. All children, whether born in or out of wedlock, shall enjoy the same social protection.[52]

Although this text is frequently considered too broad and vague, the importance of the 1948 Declaration as a universally known and recognized human rights document cannot be underestimated. It remains an important instrument to address states, which have not signed or ratified human rights treaties.[53] Moreover, in the light of the UDHR, a further UNGA resolution—the 1969 Declaration on Social Progress and Development[54]—sets forth the importance of the achievement of *'the highest standards of health and the provision of health protection for the entire population, if possible free of charge.'*[55]

Another recent and important text in this field is the Commission on Human Rights resolution entitled 'The Right of Everyone to the Enjoyment of the Highest Attainable Standard of Physical and Mental Health' (2002). The Commission, in the goal of promoting and protecting the progressive realization of the right of everyone to the enjoyment of the highest attainable standard of physical and mental health:

- Urges States to take steps, to the maximum of their available resources, to achieve progressively the attainable standard of physical and mental health by all appropriate means;

- Calls upon the international community to continue to assist the developing countries in promoting the full realization of such a right, including through financial and technical support as well as training of personnel; and
- Calls upon States to guarantee that the right to the enjoyment of the highest attainable standard of physical and mental health will be exercised without discrimination of any kind.

Secondly, the evidence that could lead to the conclusion of the customary character of the right to health is the wide participation in multilateral treaties that deal with the right to health.[56] It is submitted that the quasi-universality of participation in these treaty regimes is a strong indicator of the necessary *opinio juris* concerning the right to health, and therefore, for the underlying customary rule. As pointed out by Mark E Villiger, the international community regards the conventions as so essential for its organization that, at this stage, even if the rules did no exist *qua* contractual obligation, they would have to do so *qua* customary international law.[57]

Thirdly, we must consider the reference to the right to health in national legislation as an evidence of State practice in this field. Many constitutions around the world include the right to health within the framework of individual rights and guarantees.[58] It is, therefore, submitted that the fact of right to health is codified in many constitutions implies that States generally recognize their responsibility regarding the health of their people; and if States recognize such a right at the constitutional level, they will also logically support the existence of an international human right to health.

Lastly, further evidence of the customary character of the right to health in international law is to be found in the implementation of such a right before municipal courts.[59] Before proceeding to the analysis of the role of the judiciary in the enforcement of environmental rights and duties for the promotion of public health, it is submitted that the right to health being rooted in international customary law, as established above, India is indeed under an obligation to uphold the same and need not wait for the realization of the same under any specific treaty or Convention.

## V. An Appraisal

Several reasons may be identified as to why so few human rights and public health treaties contain express environmental rights. First, many of these treaties pre-date the development of the international community's concern with the environment. Secondly, the focus of human rights treaties differs from that of laws to protect the environment, not only because of their subject matter but also because human rights tend to be focused on an individual and anthropocentric perspective, whereas, environmental protection is more concerned with a collective ecocentric perspective. Thirdly, there are still strong doubts as to the usefulness of a public health approach to environmental protection.[60]

## VI. A Public Health Approach to Environmental Protection in India

Over the last two and a half decades, the Indian judiciary has fostered an extensive and innovative jurisprudence on environmental rights in pursuit of a better public health administration in the country. Not only has the Supreme Court ruled that every individual has a fundamental right to the *'enjoyment of pollution free water and air'*,[61] but it has been willing to resolve complex matters of environmental management according to this test, and has fashioned a series of innovative procedural remedies to accompany the new substantive right.[62] Moreover, in a country where the most serious costs of environmental damage fall upon impoverished and illiterate groups with limited access to the courts, the new environmental right and right to health is championed as a legal gateway to speedy and inexpensive legal remedy.[63]

In what follows, this section aims to identify and evaluate individual rights to environmental protection in India. Although particular attention has been paid to the new rights developed by the higher judiciary since 1985, it is impossible to assess their significance without some understanding of the larger body of Indian environmental law, including the individual rights available in other legal and statutory forms.

### VI. 1. Extra-Constitutional Rights

This paper is not the place to review Indian environmental law in its entirety, for the task is ably executed elsewhere.[64] It is, however, useful to ask why the existing regime of environmental regulation and protection was deemed so inadequate that new remedies in the form of environmental rights had to be found to foster a better public health system. Only by placing new judicially created rights against the backdrop of environmental law, as a whole, is it possible to assess their novelty and significance.

Rights may be distinguished by source and legal status. Listed in ascending order of precedence, six categories of environmental rights may be discerned in India: (i) rights derived from the customary norms of a community or caste group; (ii) rights in common law; (iii) rights derived from statutes and subordinate legislation; (iv) constitutional rights outside Fundamental Rights chapter; (v) Fundamental Rights; (vi) human rights enshrined in international instruments with legal effect in India. Leaving both types of constitutional rights aside for the moment, we may ask: *To what extent does the other four sources of legal rights provide remedies for environmental harm?*

Customary norms may generate legal rights in so far as it is recognized under the Easement Act 1882,[65] but the courts are hesitant to recognize the decisions of community councils, and the expanding scope of statutory regulation has minimized the legal impact of customary norms in recent years. It is submitted that one of the main weaknesses of the environmental regime in India is the neglect of customary laws in key ecosystems such as forests, estuaries and coastal areas.[66]

Likewise, the human rights treaties enshrined in international instruments must be regarded as a subsidiary source of rights in the Indian domestic level. Although India has acceded to a number of human rights treaties, as discussed earlier, the absence of enabling legislation deprives aggrieved parties of specific treaty-based procedural remedies. In some instances, the Supreme Court has relied upon international human rights standards to interpret statutory provisions,[67] but for the most part, the judicial reading of Indian law has not put particular emphasis on international obligations; as that would then become a political question. However, the well-settled international environment norms of *'precautionary principle'*[68] and *"polluter-pay principle"*[69] have been recognized by the Indian Supreme Court[70] as a 'universal' rule to be applied to domestic polluters as well.[71]

Turning to statutory and common rights, we may consider procedural rights under criminal and civil headings. Gateways to legal redress in criminal law depend upon statutory definitions of offences, involving public nuisance[72] and failure to comply with statutory regulations. Additionally, citizens seeking to use the law of public nuisance may rely upon three distinct but concurrent set of procedural rights.

First, an individual or group may go before a magistrate to seek a conditional order for the removal of a public nuisance under Section 133 of the Code of Criminal Procedure, 1973.[73] This power is not discretionary but mandatory so that once a magistrate is presented with evidence of a public nuisance, an order for removal must be passed.[74] Since the procedural barriers are minimal, and the burden of proof is shifted on to the person directed to remove the nuisance, Section 133 provides a speedy and summary remedy in case of urgency where damages to public interest or public health is concerned.[75]

A second procedural right relating to public nuisance may be found in the civil remedy where a person may be sued for damages at the instance of a private individual who suffers *special damage* by reason of the nuisance; *i.e.* damage, apart from that which he suffers, in common with others affected by the nuisance.[76] This provision has proved marginally more popular than other tort remedies in environmental matters, perhaps due to its link to criminal law.[77]

Closely related to the action for special damages is a third legal gateway in public nuisance. Indian law goes beyond its English progenitor in offering civil remedy even where special damage has not been sustained. Under Section 91 of the Code of Civil Procedure, 1908, two or more persons, with the leave of the court, may institute a suit for declaration, injunction, or similar relief in the case of a public nuisance.[78] The procedural requirements under this Section are not onerous, and it should provide a speedy and effective remedy for parties seeking to restrain environmental degradation amounting to a public nuisance.[79] Yet, even though Section 91 provided a channel for public interest action as early as 1908, it has been used seldom,[80] if at all used, for vindicating rights to environmental protection.

Apart from the public nuisance remedies, citizens enjoy procedural rights under the Air[81], Water[82] and Environment[83] Acts. These acts operate principally by setting emission standards, regulating the location of industry, and administering permits to pollute. All three Acts now provide[84] that any person may initiate criminal prosecution for non-compliance with statutory standards, provided 60 days prior notice is given to the Central Government. This procedural avenue is attractive because the stiffer penalties under these Acts are likely to have a much greater deterrent effect than the public nuisance provisions. As submitted earlier, this paper does not purport to scrutinize the individual environmental legislations, but it might be noted that the existing literature[85] on the same is highly critical as regards the implementation and bureaucratic functioning under the said Acts.[86]

The individual rights available in civil procedures appear more promising at first glance. Not only may tortious claims be brought up in nuisance, trespass, or negligence, but the Apex Court, in *M C Mehta v Union of India,*[87] has got beyond the *strict liability* standard of *Rylands v Fletcher*[88] in adopting the principle of *absolute liability.* The law of tort also has well-established protocols for assessing causation and for the award of damages based upon pecuniary and non-pecuniary loss—important tools in securing relief in potentially complex environmental legislation. Moreover, procedural rights are not limited to individual claims, since Order 1, Rule 8 of the Code of Civil Procedure 1908 expressly allows for *class action suits* where numerous members of a class have the same interest in the suit.[89] The procedure for class action is particularly well suited to environmental claims where a large number of people may pool their financial and institutional resources to bring actions against large companies.[90]

Given these procedural advantages, civil proceedings must be regarded as an important legal gateway to environmental justice. Ironically, however, apart from a few rare exceptions,[91] for the most part, there have been very few environmental actions in civil procedure, and most of those, which have been successful, relate to public nuisance.[92]

Thus, we arrive at an enigma of sorts: There are ample extra-constitutional rights, but they are rarely used. Admittedly, there are procedural impediments—

such as the *ad valorem* fee in tort litigation and the 60-day notice required for private prosecutions under the special statutes—which help to explain minimal citizen action. Yet certain procedural gateways, including the facility for class action suits and injunctive relief for public nuisance, are noticeably underused given the advantages they offer to aggrieved parties. It is, therefore, tempting to suspect that potential litigants are hesitant to approach the courts, and in some cases that may be true, but at least one careful study suggests that there is no lack of demand for legal remedy.[93] The more satisfactory hypothesis is that environmental and public health activists have sought relief in administrative and constitutional law.

## VI.2 Rights and Remedies in Public Law

There can be no doubt that environmental actions gravitate to the public law jurisdictions of the higher courts. Four constitutional powers are particularly relevant here. First, under Article 13, the Supreme Court is granted power to judicially review legislation, so that Acts inconsistent with the Fundamental Rights may be held *ultra vires* in the Constitution. Secondly, Article 32 guarantees individuals the right 'to move the Supreme Court by appropriate proceedings' for the enforcement of fundamental rights. Thirdly, the High Courts are also granted powers to issue prerogative writs or orders under Article 226.[94] Fourthly, Article 136 gives the Supreme Court a discretionary power to grant special leave from any judicial order, judgment or decree.[95]

A considerable power, as yet untapped for environmental legislation, is the judicial review of legislation. No statute has yet been held unconstitutional on the grounds of violating environmental rights, but this must be a realistic prospect now that the Supreme Court has recognized environmental rights as part of the fundamental rights.[96] Yet, a closely related issue had been raised, after the *Bhopal Gas disaster,* when the *parens patriae* power of the State was challenged as violative of fundamental rights on the grounds that claimants were denied access to individual legal representation under the Bhopal Gas Leak Disaster (Processing of Claims) Act 1985.[97]

## VII. Environmental and Public Health Rights under the Constitution

### VII.1 Constitutional Provisions

Article 48-A of the Constitution decrees that:

"State shall endeavor to protect and improve the environment and to safeguard the forests and wildlife of the country."

However, being a Directive principle, under Article 37, it is not justifiable and, thus, cannot be enforced in a court of law.[98]

Further, Article 51-A entails that it 'shall be the duty of every citizen...(g) to protect and improve the natural environment including forests, lakes, rivers and wildlife, and to have compassion for all living creatures.' Although the legal effect of Fundamental Duties is not spelled out, nevertheless, the duty to protect and improve the natural environment has played an important role in the emergence of an environmental right, as will be evident from the succeeding sections.

### VII.2 Expansion of Right to Life under Article 21

Following the lifting of the state of emergency in 1977, the Courts assumed a much more liberal and activist approach and were found willing to read the Directive Principles into the Fundamental Rights.[99] Even where clear obligations under Part IV did not exist, the courts have been willing to expand the scope of the fundamental rights, particularly the right to life under Article 21.[100] In *Francis Coralie Mullin v Union Territory of Delhi*[101], Honorable Supreme Court held that the right to life was not limited to the right to *'mere animal existence'*, but included a prohibition upon torture and the protection of every limb and human faculty.[102] Having set out these negative injunctions upon the State, Bhagwati J, went on to elucidate positive entitlements within the ambit of the right to life. He found that *'the right to life includes the right to live with human dignity and all that goes along with it'*.[103]

Following the *Francis Coralie* dictum,[104] two varieties of judicial reasoning may be discerned in the expansion of Article 21. One approach identifies immediate threats to survival, such as the lack of potable water, and seeks to prevent or abate such threats. A more ambitious line of reasoning proceeds where immediate survival is not threatened, but where the Court concerns itself with the quality of life. Thus, amenities are added to the baseline of physical existence, and treated as justifiable entitlements. Both of these approaches are evident in the environment cases.

### VII.3 A Constitutional Right to Environment

When the first environmental case to be entertained under the fundamental rights jurisdiction was decided, the Supreme Court gave judgment without explicit reference to the right being enforced.[105] However, the Court alluded to the *'right of people to live in a healthy environment with minimal disturbance of the ecological balance'.*[106] The legal source of this right was not discussed, although in a subsequent ruling, the court reminded 'every Indian citizen's fundamental duty' is to preserve the environment.[107] Having provided a remedy without an attendant fundamental right, the Andhra Pradesh High Court held that the *Doon Valley* judgment[108] must be construed as addressing rights under Article 21 since affecting public health caused by environmental pollution and spoliation should also be treated as amounting to violation of Article 21.'[109]

While High Courts[110] created environmental rights, the Supreme Court hesitated, and continued to issue orders under Article 32 without specifying which fundamental right was being enforced.[111] Finally, the link between environmental quality and the right to life was made explicit by the Supreme Court in the *Subhash Kumar case*[112] followed by *Virender Gaur v State of Haryana*[113] where the court observed: "Environmental, ecological, air, water pollution, etc. should be regarded as amounting to violation of Article 21. Therefore, hygienic environment is an integral facet of right to healthy life."[114]

### VII.3A. Content of the Right

The precise content of this new right is not clear. One view is that it relates exclusively to human health and, thus, amounts to a right to freedom from environmental conditions leading to death, injury or disease. In the *Shriram case,*[115] the Supreme Court found a violation of the right to life where an escape of oleum gas led to immediate injury; thus, having a direct effect upon the lives of those exposed. Even though it was a case involving industrial pollution, no environmental rights were invoked by the Court since a simple nexus between injury and exposure to the gas existed. The more difficult question is whether a court might act on a right to life or public health argument where an immediate injury is not apparent.[116] The right to a 'healthy environment' has also been invoked even where a direct nexus with human health has not been demonstrated at all.[117]

Another view is that the right relates principally to pollution rather than health. In *Charan Lal Sahu case,*[118] the Supreme Court referred to the right to 'pollution free air and water.' Whether this amounts to a negative injunction upon the state to refrain from polluting or a positive injunction to ensure supply of pollution free air and water is not entirely clear. A further narrowing of this right may be inferred from *Subhash Kumar v Bihar.*[119] Here the Supreme Court held that the right to move the court for 'removing the pollution of water or air' would lie if anything 'endangers or impairs the quality of life in derogation of laws.'[120] On an extreme view, this might be read to reduce the constitutional right to a procedural enforcement of statutory standards. This may be read as a logical extension of the public health right argument, i.e., for the complete realization of the right to health and life. It is submitted that resorting to either of the interpretations would lead to setting of the environmental standards and serve the cause, i.e., promotion of public health, just as well.

### VII.3B. Applicability to Private Entities

In an economy with a large public sector, the wide definition of State, as interpreted under Article 12,[121] permit writs to be issued against any major industries and undertakings, including those that would lie in the private sector in many economies.[122] Even where the cause of environmental harm emanates from a private entity, a writ petition against the state may be successful if the court finds that a local authority has not taken sufficient action to protect a healthy environment for local residents.[123]

### VII.3C. Conflict with other Rights

It is conceivable that a right to environmental protection could conflict with the right to religious freedom[124] or with the rights of minorities[125] if environmentally degrading practices were essential to religious practice or minority culture. More likely, however, is a conflict with the right 'to practice any profession or to carry-on any occupation, trade or business.'[126] This right is, however, subject to reasonable restrictions in the public interest,[127] a provision which appears to underlie the decision in the *Rajkot Textile.*[128]

### VII.3D. Remedies without Rights

In a number of cases, the writ jurisdictions have been used to provide equitable remedies without invoking correlative rights.[129] In cases like *Doon Valley,*[130] where no fundamental rights are discussed, detailed orders may be given without any ratio, but are apparently based on considerations of equity and technical advice.[131] More startling are orders under Article 32 requiring the Central Government to direct all educational institutions throughout India to set aside one hour per week for environmental education,[132] and that all cinemas and video parlours exhibit environmental films free of cost.[133] It is, however, difficult to imagine how such broad policy matters, applicable to the entire nation and seemingly legislative in function, have anything to do with the enforcement of an individual's fundamental right under the Constitution.[134]

### VII.4. Right to Health under the Indian Constitution

*Health is a right, not a governmental grace or service to be purchased.*[135] The Indian Constitution, reading Article 21 with Articles 39 (e), 41, 42, 47 and 51(c) casts high priority burden on the state to ensure the protection and promotion of heath of all citizens, particularly women, children and the marginalized sections of the society. The Supreme Court while broadening the ambit of Article 21 evolved the right to health as a fundamental right and the rhetoric of right to health as a human right has been translated into specific measures in jural terms.[136]

In *Vincent Panikurlanga v Union of India,*[137] the Hohorable court held that 'A healthy body is the very foundation for all human activities and hence, it is the obligation of the state to ensure the creation and the sustaining of conditions congenial to good health. It is the constitutional obligation of the State to promote health.'[138]

It was held in *CESC Ltd. v Subhash Chandra Bose*[139] that the term health implies more than mere absence of sickness. It was further held that the health and strength of a worker is an integral facet of right to life under Article 21[140] and is a fundamental human right to workmen.'[141]

In *Municipal, Ratlam v Vardichan*[142] the court emphatically noted that industries cannot make profit at the expense of public health.[143]

The Supreme Court recently prohibited smoking in public places holding that passive smoking is indirect deprivation of life and hence, violative of Article 21.[144] It is, however, submitted that such verdicts, no matter how laudable they might be, are virtually impractical to bring into actual practice and implement the same.[145]

## VIII. In Search of the Fundamental Principles

A public health approach has led the Supreme Court to derive, adopt and apply a range of principles and standards to guide the development of environmental jurisprudence. Notable amongst the *fundamental norms* recognized by the court are:

- Every person enjoys the right to a wholesome environment, which is a facet of the right to life guaranteed under Article 21 of the Constitution of India.[146]
- Enforcement agencies are under an obligation to strictly enforce environmental laws.[147]
- Governmental agencies may not plead non-availability of funds, inadequacy of staff or other insufficiencies to justify the non-performance of their obligations under environmental laws.[148]
- The *'polluter pay'* principle, which is a part of the basic environmental law of the land, requires that a polluter bear the remedial or the clean-up costs as well as the amounts payable to compensate the victims of pollution.[149]
- The *'precautionary principle'* requires Government authorities to anticipate, prevent and attack the causes of environmental pollution. This principle also imposes the onus of proof on the developer or industrialist to show that his or her action is environmentally benign.[150]
- Government development agencies entrusted with decision-making, ought to give due regard to ecological factors including—(a) the environmental policy of the Central and State Government; (b) the sustainable development and utilization of natural resources; and (c) the obligation of the present

generation to preserve natural resources and pass on to future generations an environment as intact as the one we inherited from our previous generations.[151]

- Stringent action ought to be taken against contumacious defaulters and persons who carry on industrial or development activity or profit without regard to environmental laws.[152]
- The power conferred under an environmental may be exercised only to advance environmental protection and not for a purpose that would defeat the object of the law.[153]
- In accordance with the doctrine of *'Public Trust'*[154], the state is the trustee of all natural resources, which are by nature, meant for public use and enjoyment.[155] These resources cannot be converted into private ownership.[156]

## IX. Conclusion

How do we evaluate these developments and proposals? What are the benefits and drawbacks of using a public health approach rather than an approach based in regulation, criminal law or the law of tort? Looking to the advantages, several are apparent. First, a public health approach is a strong claim, a claim to an absolute entitlement theoretically immune to the lobbying and trade-offs that characterizes bureaucratic decision-making.

A second advantage is that the procedural dimensions of an environmental right can provide access to justice in a way that bureaucratic regulation, or tort law simply cannot. A robust environmental right can mobilize redress where other remedies have failed. It is important in the Indian context wherein the procedural simplicity has made environmental rights highly attractive to aggrieved parties. An environmental right may serve as the ultimate 'safety net' to catch legitimate claims which have fallen through the procedural cracks of public and private laws.

Thirdly, a public health approach can provide the conceptual link to bring local, national and international issues within the same frame of legal judgment. At present, environmental damage is unequally distributed at both the national and international level, a non-discriminatory public health right standard could facilitate comparison, and foster political mobilization linking local concerns with more global issues.

Fourthly, a general expression of right can be interpreted creatively as issues and contexts change. This is evident in the Indian jurisprudence, where the right to a healthy environment held to be implicit in the right to life has been given more precise definition on a case-by-case basis.[157] Thus, definitions and trade-offs evolve gradually in the light of experience rather than needing to be defined comprehensively and rigidly in a single piece of regulatory legislation.

A number of limitations are also apparent. Firstly, a public health approach may not address the relationships of political economy which underlie much environmental damage. The causes of environmental damage—including technology choice, forms of production and distribution of the social product—will not be addressed by a right directed merely at their symptoms. Secondly, rights, especially procedural rights, may be used by affluent groups or 'cosmetic environmentalists' to protect a privileged quality of life, which may impose further environmental costs upon the dispossessed or environmentally vulnerable communities, who are, in turn, denied access to justice by poverty or lack of institutional skills. Thirdly, the expansion of right-based litigation may well displace other forms of legal remedy, such as tort law or negotiated settlements, which are better suited to environmental issues. This danger is identified in the Indian context where writ petitions are now displacing statutory remedies and civil suits as the main means of distributing environmental benefits and burdens. This raises the twin dangers of inconsistent standards and the transfer of essentially bureaucratic functions to the courts.

On balance, the submissions made in course of the paper show that public health approaches to environmental protection offer many attractions, and could play a key role in fostering equitable and sustainable communities. Problems of cross-cultural application, varying socio-economic circumstances, and fundamental differences in justiciability and application, all, point to the conclusion that international public health law may be best confined to a general and supervisory role, while rights developed in national law may well evolve into crucial tools for everyday environmental management, as the Indian case clearly shows. If, however, very real problems of theory and practice remain, they should stimulate careful analysis and jurisprudential innovation rather than intellectual surrender.

*(Krishnayan Sen, National University of Juridical Sciences, Kolkata. This article was prepared under the supervision of Prof. Joydasguptha.)*

## Endnotes

1 Declaration on the Human Environment, Principle 1, *Report of the UN Conference on the Human Environment (NewYork,* 1973), UN Doc. A/CONF. 48/14/Rev.1; Also see SOHN, *The Stockholm Declaration on the Human Environment,* 14 Harv. ILJ (1973), 451-5.

2 Declaration on Environment and Development, Principle 1, *Report of the UN Conference on Environment and Development* (New York, 1992), UN Doc. A/CONF 151/26/Rev.1.

3 See SHELTON, *International Environment Law,* (New York, 1991); THORNE, *Establishing Environment as a Human Right,* 19 Den. JILP (1991), 302.

4 1981 African Charter on Human Rights and People's Rights, Article 24, 21ILM (1982), 52; 1966 UN Covenant on economic and Social Rights, Article 12, 6 ILM (1967), 360; 1961 European Social Charter, Article 11, 28 ILM (1989), 156; 1989 Convention on the Rights of the Child, Article 24(2)(c), 28 ILM (1989), 1448.

5 UNGA Res. 45/94 (1990); 1982 World Charter for Nature, Principle 23, 23 ILM (1983), 455; 1989 Hague Declaration on the Environment, 28 ILM (1989), 1308.

6 See *Powell v UK,* ECHR, Ser.A..No. 172 (1990); UN Human Rights Committee, UN Doc. E/CN.4/Sub.2/1992/7.

7 See WEBER, *Environmental Information and the European Convention on Human Rights,* 12 HRLJ (1991), 177, RAMCHARAN (ed), *The Right to Life in International Law,* (1985).

8 For eg. By expanding the scope of Article 21 of the Indian Constitution, *infra* 112; Section 29 of the New Constitution of South Africa 1994; Also see BRANDL, *Constitutional Entrenchment of Environmental Protection: A Comparative Analysis,* 16 Harvard ELR(1992), 1.

9 See Human Rights and the Environment, 1st and 2nd Reports, UN Doc. E/CN.4/Sub.2/1992/7 and 1993/7; Also See Preliminary Report, UN Doc. E/CN.4/Sub.2/1991/8, and Final Report, UN Doc.E/CN.4/Sub.2/1994/9.

10 ibid

11 *Supra 1*

12 *Supra 2*

13 Final Report, UN Doc.E/CN.4/Sub.2/1994/9

14 *Supra 9*

15 See LEELAKRISHNAN, *Public Participation in Environmental Decision-Making,* in Leelakrishnan (ed), *Law and Environment,* Lucknow, 1992.

16 Universal Declaration on Human Rights, Article 19, 21; International Covenant on Civil and Political Rights 1966, Article 19, 25; Inter-American Convention on Human Rights 1969, Article 23.

17 *Supra 12.*

18 Elements of Principle 10 are reflected in Principle 23 of the World Charter for Nature [22ILM (1983), 455]; ECE Convention on Environmental Impact Assessment 1991, Articles 2(6) and 3(8); Biological Diversity Convention 1992, Article 14; Council of Europe Convention on Damage resulting from Activities Dangerous to the Environment 1993 and, as regards access to environmental information, in EEC Directives 90/313/EEC and 85/337/EEC.

19 *Union Carbide Corporation v. Union of India AIR* 1990 SC 273; Also see, BAXIand DHANDA(eds.), *Valiant Victims and Lethal Litigation: The Bhopal Case* (Delhi, 1990), xxiv; ABRAHAM, *The Bhopal Case and the Development of Environmental Law in India,* 40 ICLQ (1991) 334.

20 See *infra* 36; See *also* 1988 Additional Protocol to the Inter American Convention on Human Rights, Article 11; 1989 European Charter on Environment and Health; 1989 Convention on the Rights of the Child, Article 24 (2) (c); UNGA Resolution 45/94 (1990); and 1961 European Social Charter, Article 11.

21 See PATHAK, in Weiss (ed.), *Environmental Change and International Law,* p. 205 at 209; DUPUY, *The Right to Health as a Human Right,* (1979) p. 340. Cf. HANDL, in TRINDADE (ed.), *Human Rights, Sustainable Development and the Environment,* (1992), p. 117.

22 See Articles 16 and 20, Rio Convention on Biological Diversity1992; Also see Convention on Climate Change 1992, Article 4(3).

23 See Principle 7, Rio Declaration.

24 See Article 10, Montreal Protocol on Substances that Deplete the Ozone Layer 1987.

25 World Bank, *Instrument Establishing the Global Environmental Facility,* Geneva, March 16, 1994, 33 ILM (1994), 1273.

26 See BIRNIE and BOYLE, *International Law and the Environment,* (Oxford, 1992), 188-97; SHELTON, *Human Rights, Environmental Rights, and the Right to Environment,* 28 Stanford JIL (1991), 103.

27 The Convention has been in force since 1976; 1965 Convention on the Elimination of all Forms of Racial Discrimination, the 1979 Convention on the Elimination of all Forms of

Discrimination against Women, and the 1984 UN Convention against Torture, are, as their titles indicate, irrelevant to environmental rights and hence, beyond the scope of the instant paper.

28 Similar protection can also be found in the regional treaties such as Article 2 of the European Convention 1950; Article 4 of the American Convention on Human Rights, 1969.

29 MCGOLDRICK, *The Human Rights Committee,* (Oxford, 1991), 329-30.

30 ibid p. 330.

31 Communication No. 67/1980, *EHP v Canada, Human Rights Committee* (1990), 20.

32 The paper, however, would not look into the regional treaties like the European Social Charter, 1961 and the 1988 Protocol to the American Convention on Human Rights, as India is not a party to the said treaties.

33 See BROWNLIE, *Basic Documents on Human Rights,* 3rd ed., Oxford, 1992, 125.

34 For a fuller discussion of implications of Article 2(1) of the Covenant, see MERON (ed.), *Human Rights in International Law: Legal and Policy Issues* (Oxford, 1984), p. 213-17.

35 GA Res 2200 (XXI), UNGAOR, 21st Session, Supp No. 16, 49.

36 ibid.

37 See European Social Charter 1961, Part I, Part II, Article 11.

38 See Int. L Mat 156 (1989).

39 See PALIKA ABEYKOON, *Public Health and Human Rights—Perspectives and Issues,* NHRC, Regional Consultation on Public Health and Human Rights, 71-79 (2001).

40 World Health Organization, Constitution, July 22, 1946, preamble; *Also see* the 1978 Declaration of Alma-Ata; 1998 World Health Declaration (adopted by the 51st World Health Assembly).

41 UDHR, Article 25(1); UN Doc A/810 (1948).

42 167 UN Doc. A/44/49 (1989).

43 *Also See* 1965 Convention on the Elimination of all Forms of Racial Discrimination, Article 5 (e) (iv); and 1979 Convention on the Elimination of all Forms of Discrimination against Women, Article 11 (1) (f).

44 See *Article* 34 of the 1969 Vienna Convention on the Law of Treaties.

45 Article 38 of the Statute of the ICJ.

46 We will not deal here with the questions that may arise from the "persistent objector" State. See *Fisheries case,* Judgment of December 18, 1951: I.C.J. Reports 1951, p. 116.

47 *Nicaragua v USA* (Merits) ICJ Rep 1986, 14.

48 AKEHURST, *Custom as a Source of International Law,* BYIL 47 (1974-75) 1; Also see MERON, *The Continuing Role of Custom in the Formation of International Humanitarian Law,* AJIL 90 (1996), 238-49.

49 ICJ Rep 1985, 29.

50 See DANILENKO, *Theory of International Customary Law,* GYIL 31 (1988).

51 UDHR, Article 25(1); UN Doc A/810 (1948).

52 ibid.

53 See TOEBES, *The Right to Health as a Human Right in International Law,* Oxford, 1999, p. 40.

54 See Resolution 2542 (XXIV).

55 ibid.

56 See supra 36.

57 VILLIGER, Mark E, *Customary International Law and Treaties,* The Netherlands, Martinus Nijhoff, 1985, p. 159.

58 In the Americas, for example, some countries like Bolivia, Chile, Colombia, Ecuador, Guatemala, Honduras, Mexico, Panama, Paraguay, Peru, Suriname and Venezuela include a right to health in their constitutions either as a specific right or indirectly through the stipulation of State duties to protect health. Countries like Cuba and Nicaragua set forth in their constitutions a specific right to health and emphasize the State's role with regard to the provision of health care services. In Africa, such a right is dealt within the constitutions of Algeria, Burkina Faso, Congo, Egypt, Ethiopia, Guinea, Lesotho and Libya. Other countries whose constitutions impose a duty to protect health include: Finland, Hungary, India, Italy, Philippine, South Africa and Russia.

59 *Infra* 138.

60 See BOYLE, *Human Right Approaches to Environmental Protection,* Oxford, 1998, 108.

61 *Subhash Kumar v Bihar* AIR 1991 SC 420, 424.

62 *Infra* Heading VI. 1.

63 APARAJIT, *et.al., Judicial Response towards the Protection of Environment: A Critical Evaluation, Journal of Indian Legal Thought,* Vol.1, 2003, 97.

64 See ROSENCRANZ and DIVAN, *Environmental Law and Policy in India: Cases, Materials and Statutes,* Oxford, 2002; LEELAKRISHNAN, *Law and Environment, Lucknow, 1992.*

65 See, for example, SINGH, *Water Rights in India,* in Singh (ed), *Water Law in India, Delhi,* 1992, 15.

66 See, The Wildlife Protection (Amendment) Act 2002 newly inserted Sections 36-A, 36-C.

67 *Varghese v Bank of Cochin* AIR 1980 SC 470.

68 *Vellore Citizen's Welfare Forum v Union of India* AIR 1996 SC 2715.

69 *Indian Council for Enviro-Legal Action v Union of India* AIR 1996 SC 1446.

70 ibid.

71 See, Ministry of Environment and Forests, Government of India, *Policy Statement for Abatement of Pollution*, para 3.3 (February 26, 1992).

72 Section 268, Indian Penal Code, 1860; Section 3 (48), General Clauses Act 1937.

73 See *V Assinar v PK Moidenkutty* 1999 Cri U 4228.

74 *Municipal Council, Ratlam v Vardichand* AIR 1980 SC 1622.

75 1977 KLT 329; *Ajit Mehta v Rajasthan 1990* Cri LJ 1956 (Raj); Also see JOGA RAO, Use *of Criminal Law Machinery for Environment Protection*, (2001) 7 SCC (Jou) 58.

76 *Moolchand v Chhoga* AIR 1963 Raj 25; *Manilal v Ishwar* AIR 1925 Bom 367.

77 See *B Venkatappa v B Lorris* AIR 1986 AP 239; *Ram Singh v Babulal* AIR 1982 All 285.

78 *Aamina v Municipal Board.* AIR 1980 All 376.

79 *SN Prasad v Town Municipal Council* 1996 AIHC 5655 (Kant).

80 See *Sunik Verma v Raghubir Singh* AIR 1991 All 209, 210.

81 Air (Prevention and Control of Pollution) Act 1981.

82 Water (Prevention and Control of Pollution) Act 1974.

83 Environment (Protection) Act 1986.

84 Section 19, Environment Act 1986; Section 43 Air Act 1981; Section 49, of the Water Act 1974.

85 CHATTERJEE, *Implementation of Environment Protection Act—Problems and Perspectives* AIR 1994 Jou 113; BAXI, *Environment Protection Act: An Agenda for Implementation*, Delhi, 1987, 37.

86 BAXI, *Environment Laws: Limitations and Potential for Liberation*, in Bandopadhyay, *India's Environment: Crisis and Responses*, Dehra Dun, 1985.

87 AIR 1987 SC 1086.

88 L.R. 1 Ex. 265.

89 *Kodika v Velandi* AIR 1955 Mad 281.

90 *Cf. Charanlal v* UOIAIR 1990 SC 1480.

91 *MC Mehta v UOIAIR* 1987 SC 1086; *Mukesh Textile Mills v HR Subramanya Sastry* AIR 1987 Kant 87.

92 *Supra* 74.

93 DHAVAN, Litigation Explosion in India, Delhi, 1986.

94 See *Janki v Sardarnagar Municipality* AIR 1986 Guj 49; *KC Malhotra v State of MP* AIR 1994 MP 48.

95 *MC Mehta v UOI*AIR 1987 SC 1086.

96 See infra heading 7.3; *Rural Litigation v Uttar Pradesh* AIR 1985 SC 652, 656 *Charan Lal* supra 90.

97 *Supra* 90.

98 Cf. *Air India Statutory Authority v United Labor Unions* AIR 1997 SC 543.

99 See *Randhir v UOI AIR* 1982 SC 879 (Articles 39(d) and 14); *Hoskot v Maharashtra* AIR 1978 SC 597.

100 *Unni Krishnan v AP* AIR 1993 SC 2178.

101 AIR 1981 SC 746.

102 Id, 752-53, relying on *Sunil Batra v Delhi* 1978 Cri LJ 1741.

103 Id, 753.

104 Id.

105 *Rural Litigation supra* 96.

106 Id, 656.

107 *Rural Litigation v State of UP* AIR 1987 SC 359, 364.

108 Id.

109 *T Damodar Rao v Municipal Corporation of Hyderabad* AIR 1987 AP 171, 181.

110 *Madhavi v Tilakan* 1988 (2) KLT 730; *Kinkri Devi v Himachal Pradesh* AIR 1988 HP 4; *Koolwal v Rajasthan* AIR 1988 Raj 2.

111 *M C Mehta v* UOIAIR 1988 SC 1037; *M C Mehta v* UOIAIR 1988 SC 1115.

112 *Supra* 61.

113 (1995) 2 SCC 577,580.

114 Id 580.

115 *MC Mehta v* UOI(1986) 2 SCC 176.

116 See *Koolwal supra* 110.

117 *Rural Litigation Supra* 96.

118 *Supra* 90.

119 *Supra* 61.

120 Id.

121 *Raman Shetty v IAAI* AIR 1979 SC 1628; *Ajay Hasia v. K M Sehravardi* AIR 1981 SC 487.

122 See *Som Prakash v UOI* AIR 1981 SC 1628.

123 *See Koolwal supra* 116.

124 Articles 25-28, Constitution of India (hereafter COI); *Masud Alam v Commissioner of Police* AIR 1956 Cal 9.

125 Article 29, COI; *State of Bombay v Narasu Appa Mali* AIR 1952 Bom 84.

126 Article 19(1)(g), COI; See *Cooverjee v Excise Commissioner, Ajmer* AIR 1954 SC 220.

127 Article 19 (6), COI; *Ram Lal v Mustafabad Oil Factory* AIR 1968 P&H 399.

128 *Abhilash Textile v Rajkot Municipal Corporation* AIR 1988 Guj 57.

129 See *People United for a Better Living in Calcutta v State of West Bengal AIR* 1993 Cal 215; Janki supra 94.

130 *Rural Litigation Supra* 96.

131 See, CUNNINGHAM, *Public Interest Litigation in the Indian Supreme Court: A study in the light of the American Experience,* 29 JILI (1987), 494, 511-12.

132 *MC Mehta v* UOI (1988) 1 SCC 471, 491.

133 *MC Mehta v UOI* AIR 1992 SC 382.

134 See *Kholamuhana Society v State of Orissa* AIR 1994 Ori 191.

135 KRISHNA IYER, *The Dialects and Dynamics of Human Right in India,* Calcutta, 1999, p. 308.

136 ibid.

137 AIR 1987 SC 990.

138 Also see *Akhil Karamchari v Union of India* AIR 1981 SC 298; *PBKMS v State of W.B* AIR 1996 SC 2426; *State of Punjab VRam Lubhaiyar*(1998) Lab IC 1555.

139 (1992) 2 SCC 441.

140 Id. 462, para 30.

141 Id. 463, para 32.

142 (1980) 4 SCC 162.

143 See *MC Mehta v* UOI (1987) 4 SCC 463.

144 *Murli S Deora v Union of India* (2001) 8 SCC 765.

145 See *Kirloskar Bros. v ESIC AIR* 1996 SC 3261.

146 *Subhash Kumar supra* 61, 424; *MC Mehta v UOI* (1992) 3 SCC 256, 257; *Virender Gaur supra* 112, 581; Also see, ROSENCRANZ, *et.al, Citizens' Right to a Healthful Environment* 25 Environmental Pollution and Law 324 (1995).

147 *ICELA v Union of India* (1996) 5 SCC 281, 294-301.

148 *Dr. B L Wadhera v UOI AIR* 1996 SC 2969, 2976.

149 *S Jagannath v Union of India* AIR 1997 SC 811; *M C Mehta v Kamal Nath* (2001) 3 SCC 653.

150 *Vellore Citizens supra* 68; *A.P. Pollution Board v Prof. M V Nayadu* AIR 1999 SC 812, 819.

151 *Sate of HP v Ganesh Woood Products* AIR 1996 SC 149, 159-63.

152 *Indian Council v UOI* AIR 1996 SC 1446, 1468; *PCOHS v State of Maharashtra* AIR1991 SC 1453, 1456.

153 *Bangalore Medical Trust v BS Muddappa* AIR 1991 SC 1902.

154 *MC Mehta v Kamal Nath* (1997) 1 SCC 388.

155 *MP Rambahu v District Forest Officer AIR* 2002 AP 256.

156 Id.

157 *Virender Gaur v State of Haryana supra* 112; *Subhash Kumar v State of Bihar supra* 61; *MC Mehta v UOI* (1986) 2 SCC 176.

# 8

# Global Warming and Clean Development Mechanism Projects
## State and Trends in India*

*Jatinder S Bhatia and Harsh Bhargava*

*The issue of environmental pollution and its protection is a matter of great concern for everyone. Greenhouse Gases (GHGs) are a major source of global warming and are responsible for the depletion of the ozone layer. As a first major step to bring down the GHG concentration in the atmosphere, the Kyoto Protocol came into force last year. It has brought with itself a vast pool of Clean Development Mechanism (CDC) projects, which are currently under operation in many countries. The developing countries, including India, are maintaining an extremely positive outlook for the mechanism and exploration of new grounds to benefit from the opportunities presented. However, there is more to it than that meets the eye. It is quite early to comment on how things will shape up, in the coming years. With an emphasis on India, the authors take a look at the current project profiles, the future areas of development, the challenges faced by the project owners, and the manner in which these problems can be addressed.*

* Based on the Summer Internship Program undertaken by Jatinder S Bhatia under the supervision of Harsh Bhargava in February-May 2006, at Fieldstone Capital Services Pvt. Ltd.

*Source: The Icfai Journal of Environmental Economics, August, 2006.* 

*"Unless we protect resources and the Earth's natural capital, we shall not be able to sustain economic growth[1]".*

– Kofi Annan,
United Nations Secretary General

## Introduction

Climate change, since long time, has been debated over on a number of platforms as the world faces a gradual increase in average temperatures year after year. Rising global temperature is among one of the many challenges faced by mankind (Figure 1). However, since there is a direct link between economic development and Greenhouse Gas (GHG) emission, there was no consensus as to how to break this deadlock. It was only in December 1997, that a major breakthrough was achieved. The Kyoto Protocol is considered as a milestone in global efforts to protect the environment and achieve sustainable development, since it marked the first time that governments accepted legally binding constraints on their greenhouse gas emissions. The Protocol was ratified on February 16, 2005.

**Figure 1: Trend of Average Global Temperatures**

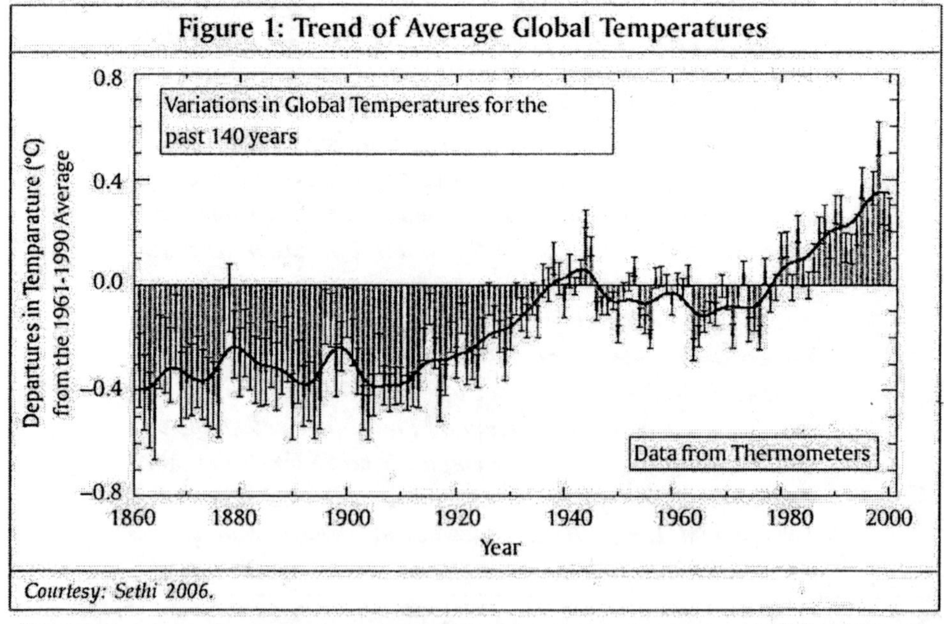


*Courtesy: Sethi 2006.*

1 Keynote address at Tufts University's Fletcher School of Law and Diplomacy, www.tufts.edu/communications/stories/052001AnnanAddress.htm, May 20, 2001.

It created a legally binding obligation for 38 industrialized countries to reduce their emissions of GHGs to an average of approximately, 5.2% below their 1990 levels over the period 2008-12 (UNEP, 2004). Although the United States and Australia are yet to ratify the Protocol, it is believed that the current framework has significant potential to bring realistic results.

The overall 5% target for developed countries to be met through reductions (from 1990 levels) varies in each nation; for e.g., it is 8% in the European Union (EU[15]), Switzerland, and most Central and East European states (UNFCCC, 2006). The Protocol does not set reduction limits on the GHG emissions of developing nations. Under the Protocol, primarily, six major gases were targeted which formed the national emission reduction strategy throughout the world (Table 1).

**Table 1: List of Potential GHGs Covered under Kyoto Protocol**

The six GHGs are not equal in terms of Global Warming Potential (GWP), which measures the relative radiative effect of GHG's compared to $CO_2$. For example, one tonne of methane has a GWP as potent as 21 tonnes of $CO_2$.

| Greenhouse Gas | GWP | Sources of Emission |
|---|---|---|
| Carbon dioxide ($CO_2$) | 1 | Fossil fuel combustion, selected industrial processes, and changes in the patterns of land-use, particularly deforestation. |
| Methane ($CH_4$) | 21 | Coal mining, landfill operations, livestock raising and natural gas/oil drilling. |
| Nitrous oxide ($N_2O$) | 310 | Fertilizer manufacturing and fossil fuel combustion. |
| Hydrofluorocarbons (HFCs) | 140-11,700 | Refrigeration and air conditioning equipment. |
| Perfluorocarbons (PFCs) | 6,500-9,200 | A variety of manufacturing processes. |
| Sulfur hexafluoride ($SF_6$) | 23,900 | Manufacturing processes where it is used as a dielectric fluid. |

*Source: UNDP 2003.*

## Clean Development Mechanism

The Kyoto Protocol (UNFCCC, 1997) envisaged to lower the overall costs of achieving the emissions targets by defining three innovative "flexibility mechanisms". These mechanisms enable countries to promote cost-effective opportunities to reduce emissions in their own country or remove carbon from the atmosphere in other countries. The three Kyoto mechanisms are:

a. Joint Implementation (JI) under Article 6 of the Kyoto Protocol provides for Annex I Parties (industrialized nations) to implement projects that reduce emissions to the atmosphere in other Annex I Parties, in return for Emission Reduction Units (ERUs).

b. Clean Development Mechanism (CDM) defined in Article 12 of the Kyoto Protocol provides for Annex I Parties to implement projects that reduce emissions in non-Annex I Parties (primarily, developing countries), or absorb carbon through afforestation or reforestation activities, in return for Certified Emission Reductions (CERs).

c. Emissions Trading, as set out in Article 17 of the Kyoto Protocol provides for Annex I Parties to acquire units from other Annex I Parties. These units may be in the form of AAUs, ERUs, or CERs.

"The CDM was designed to assist developing countries in achieving sustainable development by promoting environmentally friendly investment from industrialized country governments and businesses. It encourages developing countries to participate in the global effort to combat climate change at a time when other development priorities may limit the funding available for GHG emission reduction activities."[2]

The CDM projects help to reduce GHGs by carbon sequestration in the territory of a non-Annex I Party. The resulting CERs can then be used by the Annex I Party to help meet its emission reduction target as set under the Protocol (Figure 2). The three principle prerequisites that form eligibility criteria for a CDM project have been listed below (UNEP):

- It should lead to sustainable development in the host country,
- It should result in real, measurable and long-term benefits in terms of climate change mitigation, and
- The reductions must be additional to any that would have occurred without the project.

---

[2] *www.uneprisoe.org/cdmcapacitydev/cdmintro.pdf*

**Figure 2: Overview of the Clean Development Mechanism**

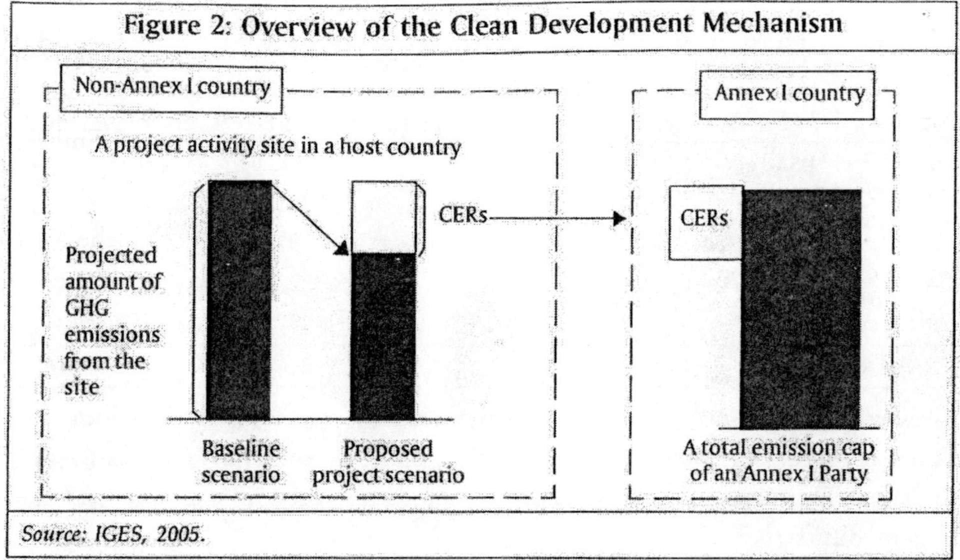


*Source: IGES, 2005.*

The following sectors may constitute a CDM project under the Protocol:

- *End-use energy efficiency improvements:* Higher efficiency refrigerators/freezers, fans/air conditioning, electric motors, etc.
- *Supply-side energy efficiency improvement:* Better transmission and distribution of heat/electricity.
- *Renewable energy:* Includes hydro power, solar thermal power, biomass gasification, wind power, etc.
- *Fuel switching:* Emission reductions by low-greenhouse emission fuels.
- *Agriculture (reduction of $CH_4$ emissions):* reduction of enteric fermentation and manure management.
- *Industrial processes:* $CO_2$ reduction from cement production, more efficient technologies (steel, paper, tobacco, etc.), reduced emission of HFCs, PFCs, and $SF_6$, etc.
- *Sinks projects:* Includes afforestation and reforestation.

## The CDM Status in India

Data furnished by the Carbondioxide Information Analysis Center (CDIAC) (Table 2) shows that India is the world's fifth largest emitter of greenhouse

**Table 2: CDIAC Rankings**

| Ranking | Country | Emission (million tonne carbon equivalent) |
|---|---|---|
| 1 | USA | 1528.8 |
| 2 | China | 761.6 |
| 3 | Russian Federation | 391.7 |
| 4 | Japan | 323.3 |
| 5 | India | 292.2 |

*Source: CDIAC, 2004.*

gases with emissions exceeding that of Germany, the United Kingdom, and Canada. Among the non-annex I countries, the $CO_2$ emissions are second only to China. This means that there exists enormous potential of emission reduction in the country.

India acceded to the Kyoto Protocol in August 2002 to fulfill the prerequisites for implementation of CDM projects, and is fast catching up with other countries to deliver a diverse portfolio of projects. In this regard, an analysis was carried out to understand the current position of India. The CDM pipeline (April 2006) issued by UNEP, Riso Center, formed the basis of the study and the key findings have been listed below[3]:

- India is on the top of the list in terms of number of ongoing projects in different host countries throughout the world. As on April 1, 2006, out of a total of 693 projects worldwide, 267 were being carried out in India. The different stages of the operations along with their CER potential have been mentioned in Table 3.
- India is also one among the leading countries in terms of number of projects registered with the UNFCCC. Out of a total 149 registered projects, 28 were from India (Figure 3).
- Out of the total 908.9 million credits expected to be generated till 2012, India is expected to contribute 20.36%, i.e., approximately, 185 million CERs (Figure 4).

[3] Summer Internship Program report titled, "Market Potential of CDM Projects and its Future Prospects, both at the International and Domestic Level" by Jatinder S Bhatia, as part of MBA program, Icfai Business School, Hyderabad, India, May 2006.

**Table 3: Status of CDM Projects in India**

| Status of Project | Number | Million CERs/year | Million CERs till 2012 |
|---|---|---|---|
| At validation (public comments for 30 days; LULUCF 45 days) | 206 | 12.2 | 101.05 |
| Request for registrátion (normal eight weeks, small-scale four weeks) | 31 | 2.3 | 18.85 |
| Request for review | 2 | 0.1 | 1.10 |
| Registered | 28 | 8 | 64.05 |
| **Total** | **267** | **22.5** | **185.05** |

*Compiled from Fenhann, 2006.*

**Figure 3: Number of Registered Projects Worldwide**

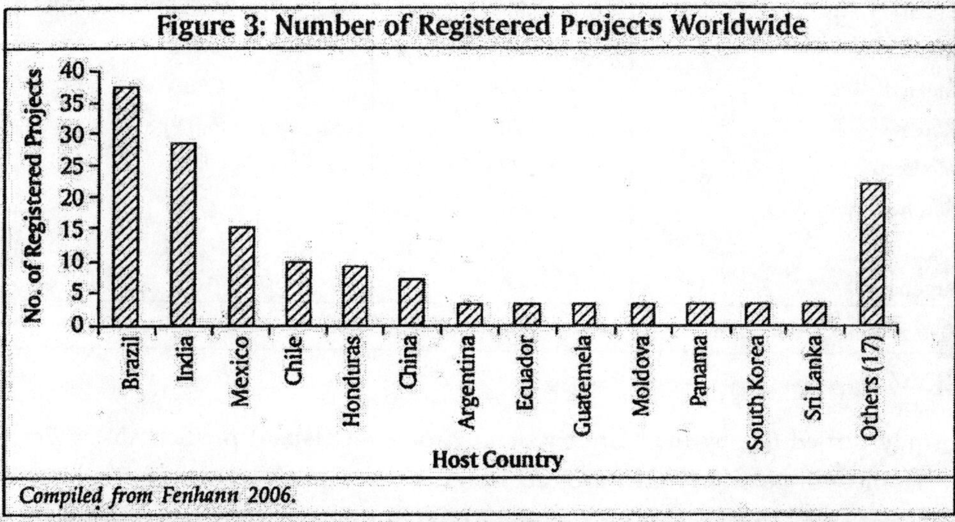


*Compiled from Fenhann 2006.*

**Figure 4: Expected % of CERs from Host Countries by the Year 2012**

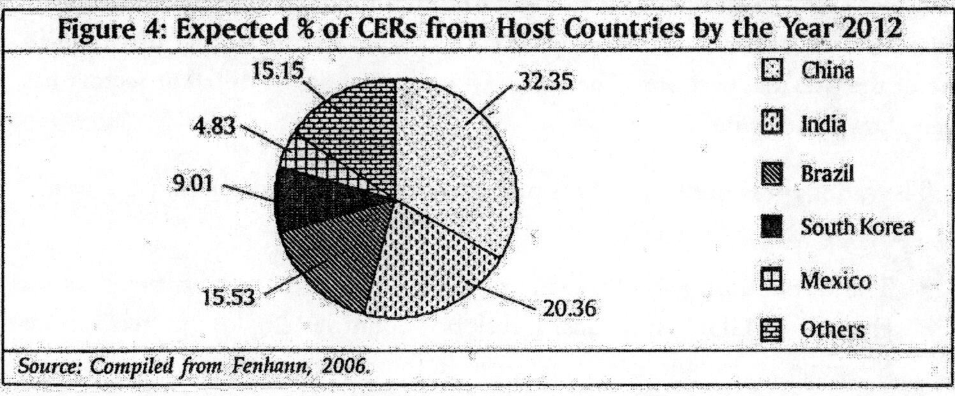


*Source: Compiled from Fenhann, 2006.*

- According to PointCarbon (Table 4), an agency providing the analysis and news for the carbon emission markets, India leads the list of countries in terms of attractiveness associated with a CDM project. This is a positive signal to a potential investor who may be interested in a CDM project in India.

**Table 4: CDM Country Ratings as on February 26, 2006 (in descending order)**

| Country | Current Rating | Rating as on December 20, 2005 |
|---|---|---|
| India | A- | (1,A-) |
| China | BBB+ | (2,BBB) |
| Brazil | BBB | (5,BB+) |
| Chile | BBB | (3,BBB) |
| Mexico | BB+ | (4,BB+) |
| Korea | BB- | (6,B+) |
| Morocco | B+ | (8,B) |
| South Africa | B | (9,B) |
| Peru | B | (7,B+) |
| Argentina | B | (10,B) |

*Source: PointCarbon, 2006.*

## CDM Opportunities in India

A study carried out by the Ministry of Environment (Japan) predicts that there are a variety of possible CDM projects in India encompassing about 300 million tonnes of $CO_2$—equivalent, including 90 million tonnes from renewable energy sources alone. Potential CDM projects cover all the sectors already discussed in one of the previous sections. The category-wise potential of different sectors has been shown in Figure 5.

The various possibilities of CDM projects in different sectors in Indian scenario are discussed below in detail:

- The disposal of waste, and the processes employed to treat these wastes give rise to GHG emissions. Developed countries employ the recovery of methane from landfills and re-use it for energy recovery to overcome methane emissions from solid waste. Such options also exist in India as

**Figure 5: Estimated Sector-wise CDM Potential in India (million CERs per annum)**

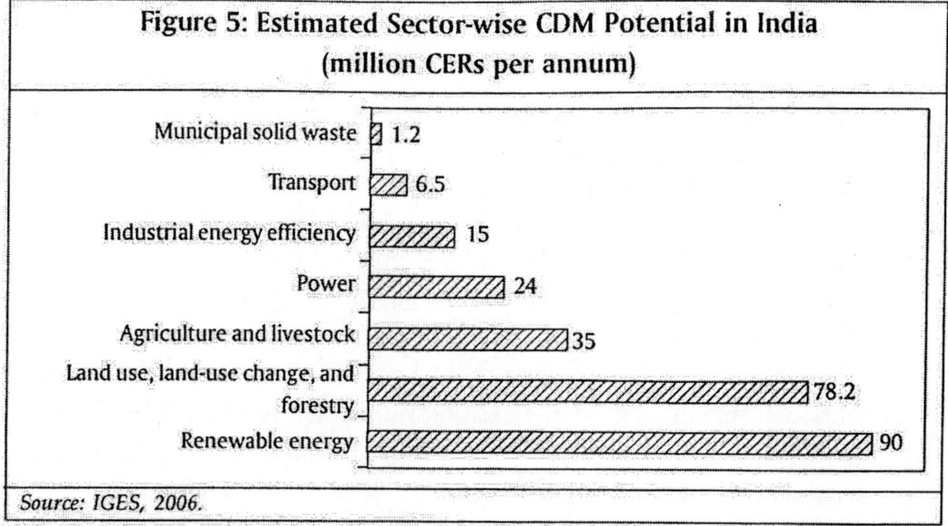


*Source: IGES, 2006.*

there are limited facilities in solid waste management and the waste is simply discarded in open dump-yards.

- Transportation is a major sector contributing to GHG emissions, which includes airways, road, rail, and waterways. Rapid urbanization is taking place in India and managing the transport sector while minimizing its ill effects like, local pollution, congestion and GHG emissions is a major challenge. The CDM bundling may be carried out by proposing biodiesel and other eco-friendly liquid fuels, and encouraging the use of Compressed Natural Gas (CNG) and Liquid Natural Gas (LNG).
- The total $CO_2$ emissions from the industrial processes in India were estimated at 99,878 Gg in 1994, out of which nearly 75%, i.e., about 75,212 Gg $CO_2$ were contributed by cement and iron and steel manufacturing processes (MOEF 2004). The major industries which look for prospects in CDM implementation (since all of them are highly power intensive) include: (i) iron and steel, (ii) cement, (iii) fertilizer, (iv) textiles, (v) aluminium, and (vi) refineries. The projects may be broadly carried out under the following categories, namely:

  (1) *Technological Enhancement:* Improving the process parameters in a sector to reduce $CO_2$ emission and heat loss.

(2) *Fuel-switching:* Use of natural gas or alternative fuels (e.g., biomass) instead of coal, petroleum coke, or fuel oil.

(3) *Recycling of Secondary Materials:* Reducing the emission of waste heat and toxic gases in the environment by reusing in the process or converting them to a less harmful form.

- In a country where around 70% of the power requirement (Figure 6) is fulfilled by fossil fuels, there exist possibilities of CDM projects as fossil fuels are the major source of GHGs. The Ministry of Power (India) has set an agenda for providing power to all by 2012. This would require an additional 46,500 MW in the existing infrastructure and hence, further chances the CDM. The improvements could be linked to better efficiency parameters leading to minimum pollution of air, water, and land.

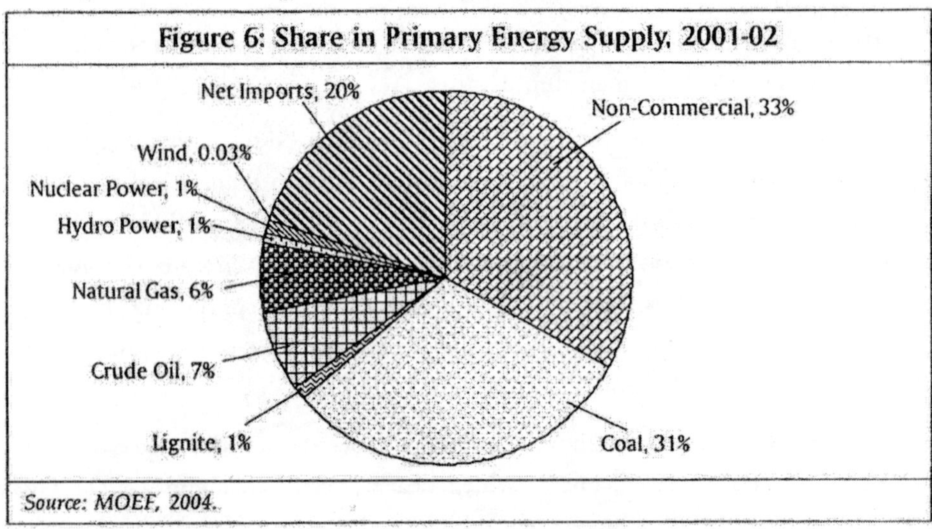


**Figure 6: Share in Primary Energy Supply, 2001-02**

*Source: MOEF, 2004.*

- The decomposition of organic animal waste in an anaerobic environment produces methane ($CH_4$). Methane produced from livestock (cattle, buffaloes, etc.) in India is the highest, among all agricultural sources, contributing about 55% of the total CH emissions (MOEF, 2004). Similarly, the anaerobic decomposition of organic material in flooded rice fields and burning of agricultural crop residue produces $CH_4$. Proper management in the agriculture and livestock sectors would also lead to major emission reductions in India.

- The forests in India, apart from providing habitat for thousands of plant and animal species, act as a sink to accommodate $CO_2$ emissions. Despite various stresses, the area covered by forests has steadily increased in India due to the proactive reforestation and afforestation programs aimed at sustainable development. If the same trend continues in the years ahead, such initiatives present considerable opportunities under the CDM regime.
- The CDM options under renewables sector are:
  - *Biofuels:* A 5-10% blending of biofuel in petro-diesel would lead to enormous savings, since it leads to "no net addition" of $CO_2$ into the atmosphere and reduces the dependence on non-renewable sources of energy.
  - *Biomass Energy:* Being an agricultural country and the world's largest producer of sugar cane, India has abundant quantities of agricultural residues and sugar cane bagasse, which can be utilized for generation of energy.
  - *Wind Energy:* The CDM can play a pivotal role supporting the growth of small-scale wind energy projects through the mobilization of capital flows. However, India has had mixed experience with wind farms set up so far with very few performing in accordance with initial projections.
  - *Hydro Power:* A large potential for small-scale hydropower plants exists in remote hilly areas in India. The development of small hydropower plants for power generation can help in rural electrification and their overall development.
  - *Solar Energy:* It represents the Earth's most abundant energy resource and can be judiciously exploited to meet ever-increasing energy requirements of the country.

## The Challenges Ahead

"One of the most challenging aspects of building a national CDM strategy is to ensure an active support from all sectors of society (civil, NGOs, private and public sector) and different sectors of the economy (industry, energy, agriculture, forestry). The private sector is essential for driving the CDM, as investors seek cost-efficient means of mitigating their emissions"[4]. But, at present, the activities in the CDM market are dominated by multilateral institutions (e.g., World Bank) and national governments, which have to meet different risk/return requirements compared to corporate players. Clearly, the low-level of engagement is due to the risk structure of CDM projects, institutional barriers, and the complexity and uncertainty in implementing CDM projects.

India can also facilitate international investment by developing a portfolio of diverse high quality CDM projects that address the needs and interests of a wide spectrum of investors. However, there are many factors acting as impediments, including inherent, country-specific problems in attracting foreign investment. Among many constraints, is the limited availability of data for informal and less organized sectors of the economy which include agriculture, forestry and many Small-scale Industries (SSI) like brick, sugar, glass and ceramic, and engineering products. In some sectors, some relevant information on CDM exists, but it has not been put together in a comprehensive form.

There are also signs of lack of preparation to host projects in India. Some of the reasons for this include premature institutional development, the existing complicated system for endorsing projects, a lack of experience among project intermediaries, and poor coordination among ministries and relevant governmental institutions.

In the Indian scenario (as shown in Table 5), it can be seen that a major proportion of CDM projects comprise biomass-based projects. However, this sector does not contribute significantly in terms of expected number of CERs. Around 27% of the projects come from the biomass sector but its contribution in annual CERs is a mere 11%. This means that even after carrying out a number of projects, India does not command a leadership position in emission credits.

It is imperative to make the situation in host countries conducive to foster CDM activities and contribute toward efforts to combat global warming. Therefore, in the Indian scenario, the two key strategies available to enhance the ability to utilize CDM are adequate dissemination of information and capacity building. This requires a sustained and timely financial and technological support, both at the individual and institutional levels.

---

[4] www.uneprisoe.org/cdmcapacitydev/cdmintro.pdf

**Table 5: Type-wise Classification of CDM Projects in India**

| Type of Project | No. of Projects | % of Projects | Annual CERs ('000) | % of CERs |
|---|---|---|---|---|
| Biogas | 12 | 4.49 | 449 | 1.99 |
| Biomass energy | 71 | 26.59 | 2507 | 11.10 |
| Cement | 19 | 7.12 | 1682 | 7.45 |
| Energy efficiency | 67 | 25.09 | 5211 | 23.06 |
| Fossil fuel switch | 14 | 5.24 | 893 | 3.95 |
| HFCs | 3 | 1.12 | 7772 | 34.40 |
| Hydro | 30 | 11.24 | 1429 | 6.33 |
| Landfill gas | 2 | 0.75 | 120 | 0.53 |
| Reforestation | 1 | 0.37 | 49 | 0.22 |
| Solar | 1 | 0.37 | 1 | 0.00 |
| Transport | 1 | 0.37 | 7 | 0.03 |
| Wind | 46 | 17.23 | 2472 | 10.94 |
| **Total** | **267** | **100%** | **22591** | **100%** |

*Compiled from Fenhann, 2006.*

## Conclusion

The whole world is trying its level best to address the issue of expected climate change in the future and its detrimental impacts. The introduction and adaptation of low emissions technologies that can support economic growth, social development and environmental sustainability is the need of the hour. In this context, the success of the CDM will be a key factor in ensuring the success of the Kyoto Protocol, and in turn, help achieve the objective of the UNFCCC. In spite of the difficulty to forecast the full extent of potential benefits available to developing countries under CDM, it is clear that its hidden potential to promote sustainable development and increase foreign investment flows is enormous. With thoughtful planning and development of a CDM strategy, India can also assist in addressing local and regional environmental problems and in advancing social goals. This way, even the developing countries would be able to play a role in climate protection. However, a long-term vision is critical to take maximum advantage of the available possibilities. India faces a host of financial, technical and development issues, which may retard the momentum of the ongoing activities. It is high time to identify ways in which CDM can generate projects with enduring development benefits and significant GHG reductions at a competitive rate in the global carbon market. This way, India can play a deciding role in preservation and protection of the environment for our future generations.

*(Jatinder S Bhatia is Student, MBA Program, the Icfai Business School, Hyderabad, India. He can be reached at jeysb@yahoo.com*

*Harsh Bhargava, Former Professor, Icfai Business School, Hyderabad. He is currently with Byrraju Foundation, Hyderabad. He can be reached at hbindia@gmail.com).*

## References

1. Carbondioxide Information Analysis Center (CDIAC) (2004), *World CO Emissions,* Washington DC.
2. Fenhann Jørgen (2006), UNEP Collaborating Center on Energy and Environment, Roskilde (*www.cd4cdm.org/publications/cdmpipeline.xls*).
3. Institute for Global Environmental Strategies (IGES) (2005), *CDM Country Guide for India,* Tokyo (*www.iges.or.jp/en/cdm/pdf/countryguide/india.pdf*).
4. Ministry of Environment and Forest (MOEF) (2004), *India's Initial National Communication to the UNFCCC,* New Delhi (*http://unfccc.int/resource/docs/natc/indnc1.pdf*).
4. PointCarbon (2006), www.pointcarbon.com
5. Sethi R K (2006), *Workshop on Clean Development Mechanism,* Ministry of Environment and Forests, India.
6. United Nations Development Program (UNDP) (2003), *The Clean Development Mechanism: A User's Guide,* New York (*www.undp.org/energy/docs/cdmchapter1.pdf*).
7. UNEP Collaborating Center on Energy and Environment (UNEP), Risø National Laboratory, Introduction to the CDM, Roskilde (*www.uneprisoe.org/ cdmcapacitydev/ cdmintro.pdf*).
8. UNEP Collaborating Center on Energy and Environment (UNEP) (2004), Risø National Laboratory, CDM Information and Guidebook, Roskilde (*www.cd4cdm.org/Publications/cdm%20guideline%202nd%20edition.pdf*).
9. United Nations Framework Convention on Climate Change (UNFCCC) (1997), *The Kyoto Protocol,* Geneva (*http://unfccc.int/resource/docs/convkp/kpeng.pdf*).
10. United Nations Framework Convention on Climate Change (UNFCCC) (2006), *A Summary of the Kyoto Protocol,* Geneva (*http://unfccc.int/essential_background/feeling_the_heat/items/2879.php*).

9

# The Socio-Technical Challenges of Safe Water Supply in Rural Bangladesh

*Wahidul K Biswas and John Merson*

***In the arsenic and salinity-affected areas of Bangladesh, rural people still struggle to procure safe drinking water due to technological, institutional and policy barriers. This paper identifies and analyzes these barriers by using a policy orientation approach. Social and decision processes are mapped to demonstrate the historic and present-day trends and conditions for water supply. Options for different potable water technologies are assessed and alternative future development scenarios using models of different water supply programs are projected for the provision of safe water for the rural communities in Bangladesh. The alternative scenarios describe a self-sustaining mechanism which will enable the poor people in rural areas to have easy access to safe water, which will further help them to secure ownership of the capital-intensive water purification technology.***

*Source: The Icfai Journal of Environmental Economics, November 2005.* 

## 1. Introduction

The shortage of potable water in rural Bangladesh is the result of lack of water conservation strategies, inaccessibility to affordable water supply options, and the weak institutional framework (Adeel, 2001). The rural poor in the inland and coastal zones suffer from a scarcity of potable water due to the presence of arsenic and salinity in the groundwater. Different options have been considered to provide safe drinking water to rural people (Azizian *et al.*, 2003; Jalil, 2003; Johnston and Heijnen, 2001; Sutherland *et al.*, 2001). Social, economical, technical, environmental and institutional factors affect the viability of these water supply options. The majority of rural people drink contaminated water, even though they know that it contains arsenic (Rammelt, 2003). Consequently, they suffer from various skin lesions, renal cancer, and gangrene in their legs, skin, lungs, bladder, and liver (SOS-arsenic.net, 2004).

Considering all these factors, this paper develops a theoretical framework for assessing appropriate potable water supply options from the socio-technical, economic, environmental, and institutional perspectives. The framework also explains how a sustainable water management plan can be developed using a policy orientation.

This paper presents the trends of water supply in both arsenic-affected villages and salinity-affected coastal areas and describes the social, technical and economic issues, including the institutional drawbacks of the present water supply system, which need to be overcome. The paper then presents an alternative approach for providing safe water through the use of appropriate water supply options as part of a sustainable water management plan for rural Bangladesh.

## 2. Assessing Appropriate Potable Water Supply Options Using a Policy Orientation Approach

A policy orientation approach as outlined by Clark (2002) is adapted here to address the safe water supply problem in rural Bangladesh.

### 2.1 Mapping Process

This process maps the past and existing water situations in rural Bangladesh, relevant to variables of social and decision processes and to describe trends in technological change, strategies, policies and the role of participations. The social process describes how different participants, with different perspectives on the safe water supply program, would undertake strategies to address the adverse water situation in the rural areas, while the decision-making process is a method or procedure to formulate different strategies to address the water situation from technical, social, economical, institutional, and legislative perspectives.

### Socio-Technical and Economic Mapping of Safe Water Options in Arsenic-Affected Villages

Because of the existence of bacterial contaminants (i.e., pathogens) in surface water, rural people have been using underground water from shallow aquifers since the 1970s (Tsushima, 2001). However, the prevalence of arsenic was observed in shallow tube well water in 1993 (Ahmed *et al.*, 2002). Both men and women from the poorest rural groups are the main victims. Women are mostly affected by Arsenicosis, as their nutritional levels are frequently deficient for reasons linked to cultural norms and reproductive functions (Curry *et al.*, 2000; WHO, 2001). Farmers who consume a large amounts of water (4-6 litres/day) with 0.2 mg/l of arsenic, i.e., 1.0 mg of arsenic per day for a few years, are vulnerable to multiple health hazards (Anwar, 2002). Higher income rural groups are able to afford better nutrition, so they are less vulnerable to arsenic contamination.

Treatment cost of arsenic-related disease is far beyond the monthly income of the rural people. The survey carried out by Barkat *et al.*, (2002) in three Bangladesh villages shows that about 73% of the rural population does not have sufficient money to bear the expenditure of the arsenic-related diseases. The average cost of treatment of arsenic symptom has been estimated as Tk.4,450, which is about 25% of the annual income of the Landless and Marginal Farmers (LMFs) which accounts for more than half of the rural population (Barkat *et al.*, 2002).

The government has initiated a program to put a green mark on shallow tube wells providing arsenic-free water (Miah, 1998). Due to the mobility of arsenic-contaminated water in the aquifer, shallow tube wells (also known as hand tube wells) with previously safe test results have later found to contain increased levels of arsenic (Harvey *et al.*, 2002; Talukder, 1999; Van *et al.*, 2002). The solution requires people to switch from the arsenic-contaminated tube wells to newly-sunk tube wells in arsenic-free zones, and this is causing socio-cultural problems (Hossain *et al.*, 2002). Poor women, who are the main managers of water for domestic purposes, cannot afford help to procure water for them. They used to walk 20 minutes a day to procure arsenic-free water; they now take 220 minutes after switching to a new arsenic-free source (Barkat *et al.*, 2002). In some instances, they are not expected to leave their cluster of houses unaccompanied, because of traditional customs.

Relatively safe levels of arsenic have been found in deep aquifers. In addition, Deep Tube Wells (DTWs), with proper maintenance, can supply large volume of water for village people throughout the year (Burgess *et al.*, 2002; McArthur *et al.*, 2004). However, there are socio-technical, economic, and geological constraints to the use of deep tube wells throughout the country. The principal barrier is the cost of sinking of deep tube wells. Although the rich may have access to safe water, socio-economic differences may impede their sharing water with the poor (Barkat *et al.*, 2002). A poor farmer has to drink arsenic-contaminated water from his own tube well, or walk a fair distance to procure water from a green-marked tube well while his rich neighbor may have deep tube wells. Such poor peasants, usually the landless and marginal farmers or LMFs, live from hand to mouth. Moreover, the male member of the household is mainly the only earning member, so the women and girls of these families have to break the cultural norm to walk for 2 to 5 hours each day to fetch water from DTWs(Salauddin and Shamim, 1994; Crow and Sultana, 2002). Bringing water home accounts for a significant portion of productive time, due to which the education of young women is adversely affected (Crow and Sultana, 2002). The burden of carrying traditional water pitchers on the hip can cause difficulties during pregnancy and deformity in posture.

A positive effect of the use of surface water, sourced from nearby ponds, canals, and rain, was observed (Das *et al.*, 2002). Arsenic toxicity in hair and nails was reduced due to drinking arsenic-free surface water. Pond Sand Filters (PSFs) can treat pond water to serve 200-500 people. This technology has not been found to be completely reliable and appropriate due to difficulties in operation and maintenance, variability of pollutant loads, conflict with fish culture, and seasonal effects (Ahmed, 2001). The cost of rainwater harvesting technologies is out of reach of middle class rural people (Ahmed, 2002). Along with seasonal supply of rain water, bacteria have been found to exist in a rainwater harvesting plant in a coastal area of Bangladesh (BRAC, 2000).

Dug wells (ultra-shallow aquifers of 1-10 m depth) are promising, and such ancient technology can provide arsenic-free water to 50 families (Crisp, 2003; Das *et al.*, 2000), although they have to be re-dug after a couple of years (Rammelt, 2005). These wells were given up because of high bacterial contamination that causes diarrhea. Such a problem can be overcome if dug wells are constructed with rounded clay slabs and walls are protected with a clay and jute layer (Smith *et al.*, 2003; SOS, 2003). However, dug wells may dry up and reduce in yield when the water table drops in the dry season or if the withdrawal due to fluctuation of demand is greater than recharge.

In sum, it appears that most of the above technologies are not completely sustainable from the socio-technical, economic and environmental perspectives in offering safe water to poor communities in Bangladesh.

**Socioeconomic Mapping of Safe Water Options in Salinity-Affected Coastal Areas**

Salinity is the main issue for drinking water in the coastal areas where 25% of the total population of Bangladesh lives (Ali, 2001). The availability of saline-free pockets in coastal areas is lower than the availability of arsenic-free pockets in the arsenic-affected rural villages, whereas in other places, neither ground nor surface water is saline-free (Rahman *et al.*, 1997). According to UNICEF and DPHE in 1994, in hilly areas and the coastal belt, over 20% of people have to travel more than 200 m to get potable tube well water. In contrast, about 85% of rural Bangladeshi households have access to safe drinking water within 150 m, even though the lowest level of salinity in a shallow tube well is about twice the WHO level, i.e., 0.5 mg/L (DoE, 1991; BUET-BIDS, 1992). Although deep tube wells provide a relatively reduced level of salinity, the water contains sand which makes deep well water undrinkable in coastal areas (Ahmed, 1996).

Ponds are the only water source in coastal areas where salinity exists at an acceptable level (Ahmed, 1996). As a result, pond sand filters have been used to treat bacterial contaminants to supply pond water for drinking purposes. However, ponds are not always available in coastal areas, and they are used for income generation from aquiculture (BRAC, 2000). Therefore, there are one to two ponds in two to four villages exclusively reserved for the supply of drinking water. As a result, people, usually women, have to wait for a long time in a queue to collect water from the PSFs and carry it along a great distance. Some of these ponds remain dry during summer. In this situation, people need to travel by boat for 4 to 5 km to collect drinking water (Azam, 1996). Water concerns are also differentiated by class in coastal areas. Richer households can generally create and maintain fresh water ponds, or they can afford to purify water for consumption. Poorer women, by contrast, have to walk considerable distances (often over 2-5 km) to obtain drinking water.

**Institutional Mapping of a Potable Water Supply Program**

There are two main types of institutions which implement water supply programs or disseminate technologies to the rural people: Government and NGOs. Some major participants of these institutions are:

- The relevant government institutions, Department of Public Health Engineering (DPHE) and the Local Government Engineering Department, are usually involved in the testing and distribution of water and in sinking tube wells.
- NGOs are regarded as implementing institutions for their ability to reach the grass roots. Some NGOs, such as the Bangladesh Rural Advancement Committee (BRAC), Proshika and Grameen Bank, have their own financing model and scheme for providing loans to small rural enterprises, while NGOs like NGO Forum, Christian Aid and Dhaka Community Hospital specifically conduct water supply programs.

Other than these organizations, there are research organizations, such as Bangladesh University of Engineering and Technology and Dhaka University which have accessed different water supply programs and have also developed the technology required to suit the situation. Some new companies have recently emerged as manufacturers or importers of water technologies.

Although the National Policy for Water Supply and Sanitation has a noble objective of facilitating access of all citizens to the basic level of services in water supply and sanitation through capacity building, credit extension, research and development activities, majority of the rural people are still suffering from the scarcity of potable water due to lack of appropriate strategies (LGRD, 1998; Rammelt, 2003; SOS-Arsenic.net, 2004). According to Milton (1999), the same report written on *"Arsenic Crisis: Issues Need to be Addressed* " in 1998 can be reproduced in 1999, because no significant change in policy and strategies, except for the technologies, took place from 1993 when arsenic was first identified in the shallow tube wells.

According to the National Policy for Water Supply and Sanitation, DPHE will act as the lead agency with the stakeholders including NGOs and private sector in providing coordinated inputs into the rural water supply projects (LGRD, 1998). For example, Bangladesh's largest safe water supply project "Bangladesh Arsenic Mitigation Water Supply Project (BAMWSP)", supported by the World Bank, has the mandate to work with DPHE. Despite DPHE's impressive network offices at the Sub-district and District levels, the department is hopelessly subjected to inefficiency, bureaucracy and corruption, amôngst other problems (Hoorens and Koender, 1999). Government's top-down approach has not been found to be suitable to reach to the level of a village (Rammelt, 2003). Most importantly, it lacks programs or tools, such as monitoring, onsite support, financing mechanisms, and adequate training required during the implementation phase of projects.

In contrast to BAMWSP, UNICEF has chosen to utilize the capacity of Bangladesh's large national NGOs (BRAC, Grameen Bank, NGO Forum, Dhaka Community Hospital, etc.) to research peoples' responses to various methods of mitigation in four districts in order to evaluate their effectiveness (Jakariya, 2003; Patel, 2001). Although NGOs have been involved in implementing the program, they did not play a role in micro-financing and capacity-building to develop a self-sustaining mechanism for the rural poor to obtain access to potable water, and to accelerate the technology dissemination process.

In addition, water-related policies so far developed over the decade have not been found to be consistent with each other. According to the National Policy for Water Supply and Sanitation, adoption of water supply technology options should be appropriate to specific regions, geological situations and social groups

(LGRD, 1998). In contrast, the National Policy for Arsenic Mitigation, prepared in 2004, has predetermined technological options without taking social, technical and geological factors into account (SOS-Arsenic.net, 2004). However, the national policy involves Local Government Institutions (LGIs) with technical support from relevant government agencies to provide safe water options and health services at the grass roots level, Previous experience shows that it is unlikely that the programs would succeed. It was observed in previous programs that the community, led by LGIs, had the least control over the safe water supply options, and the allocation of deep tube wells mostly favored the Chairman and Members of LGIs (Ahmed, 2002). The National Policy for Arsenic Mitigation gives preference to surface water over groundwater as the water supply source. However, this may not be a usual situation throughout the country due to differences in geological characteristics and water resources utilization patterns.

Bangladesh Water Supply Program (BWSP) is commencing a US$50 mn project, of which 90% is coming from the World Bank, to ensure piped water supply in large villages and growth centres and non-piped water arsenic mitigation activities in the country (BSS, 2005). Under this program, people of only 500 villages will be provided with safe water (BSS, 2005), while groundwater of 2,000 villages contain arsenic above 50 ug/L—about five times the WHO level (Rahman *et al.*, 2003). A socially equitable investment policy is, therefore, inevitable to provide safe water to the majority of the arsenic affected villages of the country.

It appears from the foregoing discussion that the ongoing water problem in Bangladesh arises from a number of technical, financial, institutional and policy factors.

## 2.2 Trend Analysis

Trends, obtained from the mapping process, ascertain whether the decision-making processes, including plans and programs so far developed, have adequately been improved to satisfy the variables of social processes or sustainable water supply systems (i.e., water demand, continuing supply, affordability, health, reduce drudgeries etc.); how technologies are changing over time, how the government policies are responding to water problems, is the awareness of rural people increasing?

Is the institutional structure sufficiently strong to channel water supply options to the actual recipients? The above section shows that these trends are unable to meet the goal. Some examples of the technological issues are green marked tube wells with uncertainties in supplying arsenic free water, dug well and rain water harvesting with seasonal water supply, deep tube wells with cost constraints and pond sand filters with their inability to cope with the variable pollutant loads, etc. For policy issues, some examples are market failure[1] for water resources due to cost sharing approach, absence of technology assessment, lack of decentralization and a top down approach. Safe water supply strategies and policies, so far developed in Bangladesh, have not taken long-term sustainability adequately into account. Some technical solutions have been attained, but the socio-cultural and economic situations have impeded the technology dissemination process. Some examples of institutional issues are a lack of integration between the government and the NGOs, lack of community participation and absence of self-sustaining implementation models. Considering all these factors, two approaches have been considered as solution to this problem: *Technology assessment* and *reviewing successful models.*

## 2.3 Alternative Solution

Two approaches to alternative solutions: *Technology assessment* that determines appropriate technology which is socially, culturally, economically, technically, environmentally and institutionally feasible, and *reviewing successful models* on relevant problems to overcome policy and institutional gaps. These two approaches are briefly discussed below.

### Technology Assessment

The relevant implementing authority, which could be either a government agency or an NGO, would subcontract either an engineering consulting firm or a university research team to carry out the techno-economic assessment processes of safe water supply options. Following the techno-economic assessment process, the implementing authority would finance as well as implement safe water supply options in collaboration with NGOs, lending organizations and training institutes. The steps that would be carried out for technology assessments are:

---

[1] The inability of markets to reflect the full social costs or benefits of a good or service, with markets not resulting in the most efficient or beneficial allocation of resources.

Firstly, existing and alternative options for delivering the same level of service have been identified and a comparative study has been undertaken. Once the options are identified, different water technologies would be designed to optimize their capacities. Such technologies would be chosen that will help the rural people to easily operate without specialized skill and technical knowledge. Technologies using locally available raw material and easily procurable resources will be given top priority, for example, hand pumps for lifting water where local people participate and the use of dug wells which do not require chemical treatments, etc. The variables of the design, such as size ($m^3$ of water purified/$m^2$ of the filter area), chemicals (kg/$m^3$ of water treated), energy (kWh/$m^3$ of water treated), and person-days (/$m^3$ of water treated) are dependent on the demand for potable water (l$m^3$/day), the level of contaminants (number of parts per million), and the size of the contaminants to be removed, while the energy requirement is a dependent variable of the capacity of the purifier. The design process may be constrained by the availability of energy resources to power water purifiers and the availability of person-days, as rural people may be engaged in other activities during the peak agricultural seasons, and a lending limit for running such projects.

Once the size and capacity of the technologies are ascertained, corresponding capital costs (e.g., \$/$m^3$ of water treated or $m^2$ of the technology) would then be found; this is required for financial analysis in the following stage. The financial viability of the above technological options is based on the estimation of parameters such as water production cost, size of investment and payback period. The project should be within the lending limit of about \$3,990 which is the maximum offered by the Bangladesh Rural Advancement Committee for small rural industries. The payback period should be two to three years which is the same payback period for loans, offered by Grameen Shakti in Bangladesh, to purchase solar home systems (Biswas *et al.*, 2004).

Following the financial analysis, a socioeconomic analysis would be carried out to overcome some general socio-cultural and institutional constraints which are:

- Poor people have limited access to a high quality water supply due to cultural restrictions which have become social norms. For example, rural women usually perform all household activities such as cooking, cleaning, bathing and feeding children, platform reinforcement and fuel collection other than fetching water (Biswas *et al.*, 2001b). Thus, the constraint on women's time and the additional work time involved in obtaining water from non-contaminated sources, are likely to encourage continued use of contaminated wells (Crow and Sultana, 2002).

- Safe water for household use is dependent on the time and labor involved in the collection of water, the time and other resources used in boiling or sterilizing the water, and in managing water within the household.
- Because of religious and customary restrictions, women are discouraged from going beyond their housing cluster.
- Gender inequality also exists, as the labor cost of collecting water is borne largely by women and girls who are responsible for domestic chores in most developing countries.
- Apart from physical drudgery related to procuring water, poor women do not have time for education, while there is also a chance of miscarriage of pregnant women from the extreme labor involved in carrying water.
- Poor people account for a significant portion of the rural population, and cannot afford to pay for technology offering year round water supply on a long-term basis.
- The survey conducted in Bangladesh shows that poor people pay someone Tk.60 per month to bring deep tube well water from a large distance (Biswas, 2003, unpublished). Although people now pay for water, there is no appropriate institutional framework to create a water market where the true cost of water is reflected, where people would be able to purchase water at affordable rates.

The proposed strategy would be able to tackle such situations through the creation of a water market run by the rural poor, who will have additional income from selling water to meet basic expenditure of life as well as a supply of safe water for themselves (Biswas *et al.*, 2001a; Biswas, 2003; Biswas *et al.*, 2004). Such a mechanism is self-sustainable as the cost of water is within the purchasing power of consumers, and the loan taken for this capital intensive technology is recovered from selling water to the rural population. Once the payback period is over, the poor people involved in the business will own the technology involved.

Through such a water business project, the government would be able to recycle the capital of one project to another project in another arsenic or salinity-affected village; thus the government's investment in safe water supply options would be equitably and efficiently used.

**Figure 1: Socioeconomic Model for Safe Water Supply in Rural Areas**

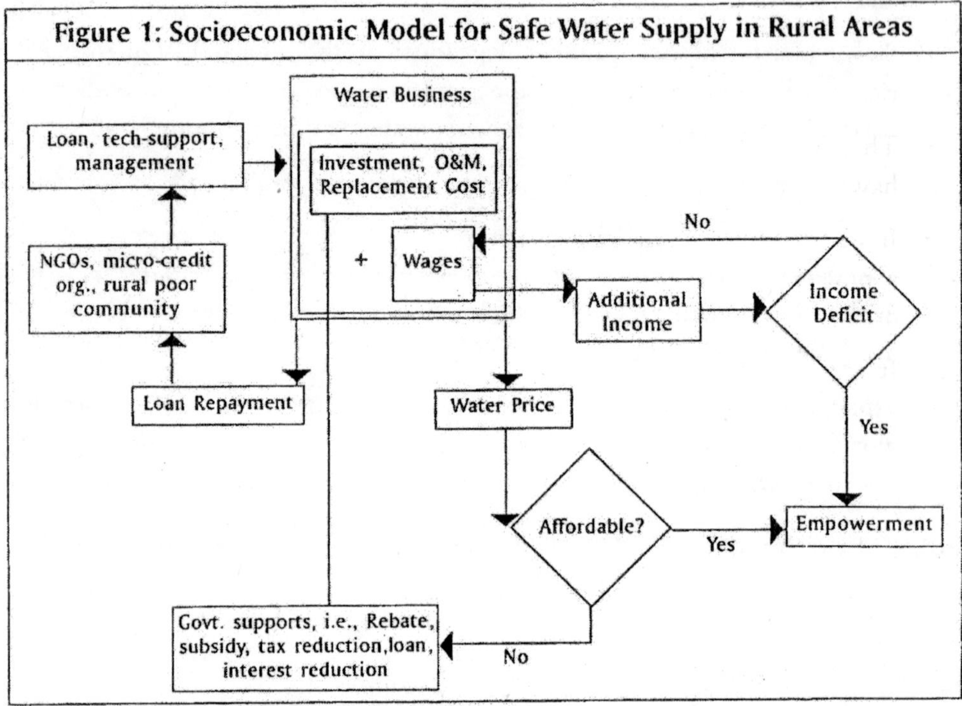


Figure 1 shows the flow diagram of the socioeconomic model with key stakeholders involved in the business. As can be seen in this figure, the implementing organization (i.e., NGO) with assistance from other facilitating organizations like micro-credit organizations, companies and technical support institutions, provides financial and technical assistance. Two important variables are prices and wages. If the price is not affordable, macro-economic policies such as tax exemption, subsidies, rebate, extension of loan, would be considered. If the wages provide insufficient income to meet the income deficit of the people involved in the business, wages would be increased until the price is affordable to the rural community. Optimum price is obtained when all people can afford to pay this rate and when wage rate of the landless and marginal farmers in the business help cover their income deficits. The following outcomes are anticipated due to the introduction of this type of business:

- The poor involved in the water business will maintain water purifiers as well as distribute water to consumers. This has a dual economic benefit. On the supply side, the people involved in the business would be paid for their time and effort in delivering water. On the demand side, the time and physical effort previously spent on procuring water would be reduced.
- Social equity would be enhanced as the poor will be less dependent on the rich for potable water.
- The situation will be in harmony with the cultural norm: Women will not have to leave their housing cluster as water would be delivered to the home.
- It will strengthen the demand-side management of water delivered. With a piped water supply, some people would have used excess water, while others would have suffered from insufficient water.
- It will also reduce the demand on natural resources.

Most importantly, poor people participating in the business will not have to contribute to the capital cost of technology and the true cost of water will be reflected in the price. It will be able to overcome the problems of the current financing system where the rural people share the cost of technology.

Complying with the outcomes of this approach will determine the best technological option in socio-technical and economic contexts. The results obtained from the technology assessment analysis, such as facilities (e.g., reduction of import tax, rebate scheme, lending limit), logistics (infrastructures, capacity building), technical processes (new innovative technological solution), and strategies (marketing, privatizing, restructuring) will be used to overcome the gaps of technology and policy, and help future development.

**Reviewing Different Successful Models**

The second task is the study of different successful models to address the policy and institutional gaps in water supply in Bangladesh. First, the literature on countries with similar socioeconomic conditions and successful implementation of water supply projects will be reviewed. Using a matrix, we will then compare their water supply programs, including technology, policies, and institutions, with those of rural Bangladesh against the variables for decision-making and social conditions mentioned earlier. The outcomes of different countries' implementation models will also be compared on the basis of technological development, accessibility of drinking water, improvement in health and hygiene, and water service facilities. This will allow us to compare appropriate policies from other countries with Bangladesh's situation.

## 2.4 Alternative Future Development Scenario

The following scenario is expected to happen due to the introduction of the above solution approaches to combat water problems:

Local people will have direct control on the water supply system. As a requirement of the criteria of the appropriate technology, local resources will be harnessed as much as possible. Thus, these technologies could be easily managed and operated by the local people. Poor people involved in the business will get ownership of the technology without requiring any capital debt.

Dependence on government support, which is a time consuming process, and is affected by bureaucracy and corruption, would be lessened. The water business concept will be able to overcome the drawbacks of the government's cost sharing approach. In the present case, the implementing organization would use the recovered capital cost of technologies in other villages to further address the contaminated water problem.

The government's investment in safe water supply options would be socially equitable and would be efficiently used. For example, the investment equivalent to US$50 mn for supplying safe water to 500 villages by the Bangladesh Water Supply Program could be recycled to safe water supply projects in other villages where the groundwater is contaminated with arsenic.

Poor women will not have to depend on the rich for potable water, while they will be able to purchase water at affordable prices and be free from physical drudgery.

Extension of water business to an electricity business for domestic and agricultural purposes[2] could happen, because 89% of the villages in Bangladesh do not have access to electricity (Marphy *et al.*, 2002). Decentralized renewable energy would be required to power some of these water purification technologies. This could reduce the unit cost of production of the individual components of this integrated business. The employment sector in the village, however, would increase. Since farm jobs are scarce due to population pressure and fragmentation of land, creation of non-farm jobs like water services would empower the people.

## 3. Conclusions and Recommendations

The rural poor in the inland and coastal zones of Bangladesh are the ultimate victims of arsenic poisoning and saline groundwater consumption. Social, economical, technical, environmental and institutional factors are all involved in failure to develop a sustainable system which can supply safe water to rural people. This paper has taken a policy orientation approach to this problem in an effort to demonstrate the causes of this failure. The following strategies are needed to address the failure of water supply programs in rural Bangladesh:

- A socially equitable model of natural resource management.
- A self-sustaining mechanism or water business concept that would enable rural people to get access to safe water at an affordable price and empower them as well through an income generating water business.
- An integration between NGOs, rural community, government and research organizations to facilitate water business in the arsenic and salinity-affected villages.
- A business to assist in recovering the loan for the capital-intensive safe water supply options.
- Local employment creation through inputs from the rural people to operate and maintain the technology, as well as to deliver water to customers' houses.

*(Wahidul K Biswas is a Postdoctoral Fellow, Research Program in Environmental Management and Policy, School of History and Philosophy of Science, The University of New South Wales. He can be reached at w.biswas@unsw.edu.au*

*John Merson is a Director, Research Program in Environmental Management and Policy, School of History and Philosophy of Science, The University of New South Wales. He can be reached at j.merson@unsw.edu.au).*

## Acknowledgement

The authors would like to thank Rosalie Chapple of UNSW, Crelis Rammelt and Jan Boes of Technology University of Delft, Abul Hasnat Milton of the University of Newcastle and Phil Lawn of Flinders University for their critical comments, and Lynne Taaffe for her editorial support.

## References

1. Adeel, Z (2001), "Policy Dimensions of the Arsenic Pollution Problem in Bangladesh", Technologies for Arsenic Removal from Drinking Water, workshop on Technologies for Arsenic Removal from Drinking Water organized by Bangladesh University of Engineering and Technology (BUET), Dhaka, Bangladesh and The United Nations University (UNU), Tokyo, Japan.
2. Ahmed M F, (2002), *Alternative Water Supply Options for Arsenic Affected Areas of Bangladesh,* International Training Network (ITN) – BUET, Dhaka – 1000.
3. Ahmed, C M (2002), "Impact of Arsenic on the Rural Poor in Bangladesh", *Bangladesh Environment 2002,* 2nd International Conference on Bangladesh Environment Bangladesh Poribesh Andolon (BAPA), Dhaka.
4. Ahmed, K M, Van Geen, A, Zheng, Y, Stute, M, Shamsudduha, M, Dhar, R, Horneman, A, Steckler, M, Versteeg, R, Gavrieli, I, Seddique, A A, Aziz Z and Hoque, M A (2002), "Arsenic in Groundwater of Araihazar: Occurrence, Distribution and Mitigation", *Bangladesh Environment 2002,* 2nd International Conference on Bangladesh Environment Bangladesh Poribesh Andolon (BAPA), Dhaka.
5. Ahmed, M F (1996), "Coastal Water Supply in Bangladesh, Reaching the Unreached: Challenges for the 21st Century", 22nd WEDC Conference New Delhi, India.
6. Ahmed, M F (2001), "An Overview of Arsenic Removal Technologies in Bangladesh and India", *Technologies for Arsenic Removal from Drinking Water,* workshop on Technologies for Arsenic Removal from Drinking Water organized by Bangladesh University of Engineering and Technology (BUET), Dhaka, Bangladesh and The United Nations University (UNU), Tokyo, Japan.
7. Ali, A, (2001), "Vulnerability of Bangladesh coastal region to climate change with adaptation options", Bangladesh Space Research and Remote Sensing Organization (SPARRSO) Agargaon, Sher-e-Bangla Nagar, Dhaka 1207, Bangladesh.
8. Ali, M and Tarafdar, S A, (2002), "Drinking Water Arsenic Pollution in Bangladesh and

Health Effects", *Bangladesh Environment 2002,* 2nd International Conference on Bangladesh Environment Bangladesh Poribesh Andolon (BAPA), Dhaka.

9. Anwar, H M, Akai J, Mostafa, K M G, Saifullah, S and Tareq, S M, (2002), "Arsenic Poisoning in Groundwater: Health Risk and Geochemical Sources in Bangladesh", *Environment International* 27, pp. 597-604.

10. Azam, K A, (1996), *Mass Communication and Social Revolution for Development: Environment and Sanitation,* NGO Forum, Lalmatia, Dhaka.

11. Azizian, F, Evans, K and Afzal, M S, (2002), "Use of Activated Alumina-based Adsorption Media for Arsenic Remediation in Bangladesh", *Bangladesh Environment 2002,* 2nd International Conference on Bangladesh Environment Bangladesh Poribesh Andolon (BAPA), Dhaka.

12. Bangladesh Rural Advancement Committee (BRAC), (2000), *Combating a Deadly Menace, Research and Evaluation Division,* BRAC, 75 Mohakhali Dhaka 1212.

13. Barkat, A, Maksud, K M, Anwar, K S and Munir, A K M, (2002), "Social and Economic Consequences of Arsenicosis in Bangladesh", *Bangladesh Environment 2002,* 2nd International Conference on Bangladesh Environment Bangladesh Poribesh Andolon (BAPA), Dhaka.

14. Biswas, W K (2003), "Dissemination of Biogas Technologies in Rural Bangladesh: A Case Study", in *Bangladesh's Development Agenda and Vision 2020: Rhetoric or Reality?* University Press Limited, Dhaka.

15. Biswas, W K, (2003), *Field Survey Conducted in Arua and Moheswarpur, Dighalia, Khulna,* Environmental Science Discipline, Khulna University, Bangladesh, Khulna.

16. Biswas, W K, Bryce, D, and Bryce P (2001a), "Technology in Context for Rural Bangladesh: The Options from an Improved Cooking Stove for Women", *ISES 2001 World Solar Congress,* pp. 1441-1451.

17. Biswas, W K, Bryce, P and Diesendorf, M, (2001b). "Model for Empowering Rural Poor through Renewable Energy in Bangladesh", *Environmental Science and Policy, 4(6),* pp. 333-344.

18. Biswas, W K, Diesendorf, M and Bryce P (2004), "Can Photovoltaic Technologies Help Attain Sustainable Rural Development in Bangladesh", *Energy Policy,* 32(10), pp. 1199-1208.

19. BSS (Bangladesh Sangbad Sangstha), (2005), "Tk 319.82 Crore Program to Ensure Piped-water Supply in Villages Launched", National News Agency of Bangladesh, 4 June.

20. BUET-BIDS, (1992), *Multipurpose Cyclone Shelter Program,* Final Report, UNDP/ World Bank Project BGD/91/025.

21. Burgess, WG, Burren M, Perrin, J and Ahmed K M (2002), "Constraints on Sustainable Development of Arsenic-bearing Aquifers in Southern Bangladesh", Part 1: A Conceptual Model

of Arsenic in the Aquifer. In: *Sustainable Groundwater Development,* Eds. Hiscock K M, Rivett M O and Davison R M, Geological Society, London, Special Publications, 193, 145-163.

22. Chowdhury, A N (1989), *Let Grassroots Speak,* The University Press Limited, Dhaka.
23. Clark, T W (2002), *The Policy Process: A Practical Guide for Natural Resource Professionals,* Yale University Press, US.
24. Crisp, P T, (2003), *Personal Communication, School of Chemical Engineering and Industrial Chemistry,* UNSW, Sydney NSW 2052.
25. Crow, B and Sultana, F (2002), "Gender, Class and Access to Water: Three Cases in a Poor and Crowded Delta", *Society and Natural Resources,* 15, pp. 709-724.
26. Curry, A, Carrin, G, Bartram, J, Yamamura, S, Heijnen H, Sims, J, Hueb, J and Sato, Y, (2000), *Towards an Assessment of the Socio-economic Impact of Arsenic Poisoning in Bangladesh,* World Health Organization, WHO/SDE/WSH/00.4.
27. Das, H K, Sengupta, P K, Hossain, A, Islam, M and Islam, F, (2002), "Diversity of Environmental Arsenic Pollution in Bangladesh", *Bangladesh Environment 2002,* 2nd International Conference on Bangladesh Environment Bangladesh Poribesh Andolon (BAPA), Dhaka.
28. Das, R, Bidyut, J Y, Shohel, A E, Protik, T H, Islam, S, Uddin, M, Islam, A, (2000), "Are Dugwells a Good Option for the Community?", 3rd International Arsenic Conference, Dhaka Community Hospital, Dhaka, Bangladesh, May 13-14.
29. DoE (Department of Environment), (1991), *Environmental Quality Standards (EQS) of Bangladesh.* Government of the People's Republic of Bangladesh.
30. Harvey, C F, Swartz, C H, Badruzzaman, ABM, Koen-Blute, N, Yu, W, Ali, M A, Lay, J, Beckie, R, Volker, N, Brabander, D, Islam, S, Ahmed, F, (2002), "Arsenic Mobility and Ground Water Extraction in Bangladesh", *Science,* 298, pp. 1602-1606.
31. Hoorens, S and Koenders, P (1999), "Together Against Arsenic Aggression", *The Daily Star,* popular newspaper, November 26.
32. Hossain, M Z, Gani, M S, Jakariya, M and Chowdhury, AMR, (2002), "People do Switch to Arsenic-free Water", *Bangladesh Environment 2002,* 2nd International Conference on Bangladesh Environment Bangladesh Poribesh Andolon (BAPA), Dhaka.
33. Jakariya, M, (2003), *The Use of Alternative Safe Water Options to Mitigate the Arsenic Problem in Bangladesh: Community Perspective,* Research Monograph Series No. 24, Research and Evaluation Division, Bangladesh Rural Advancement Committee, Mohakhali, Dhaka.
34. Jalil, M A, (2002), "Removing Arsenic from Drinking Water by Reverse Osmosis", *Bangladesh Environment 2002,* 2nd International Conference on Bangladesh Environment Bangladesh Poribesh Andolon (BAPA), Dhaka.

35. Johnston, R and Heijnen H, (2001), "Safe Water Technology for Arsenic Removal", Technologies for Arsenic Removal from Drinking Water, workshop on Technologies for Arsenic Removal from Drinking Water organized byBangladesh University of Engineering and Technology (BUET), Dhaka, Bangladesh and The United Nations University (UNU), Tokyo, Japan.

36. LGRD (Local Government Rural Development and Cooperatives), (1998), "National Policy for Safe Water and Sanitation", LGRD, Government of the People's Republic of Bangladesh.

37. Marphy, R, Kamal, N, and Richards, J R, (2002), *Electricity for All,* Bangladesh Rural Electrification Board, Dhaka.

38. McArthur, J M, Banerjee, D M, Hudson-Edwards, K A, Mishra, R, Purohit, R, Ravenscroft, P, Cronin, A, Howarth, R J, Chatterjee, A, Talukder, T, Lowry, D, Houghton, S and Chadha D K (2004), "Natural Organic Matter in Sedimentary Basins and its Relation to Arsenic in Anoxic Ground water: The Example of West Bengal and its Worldwide Implications", *Applied Geochemistry,* 19 pp. 1255-1293.

39. Miah, F, (1998). "In Quest of Safe Water for Rural Bangladesh", Department of Public Health Engineering, Bangladesh.

40. Milton, A H (1999), "Stakeholders of the Arsenic Trade", *The Daily Star,* Dhaka, August 20.

41. NGOF (NGO Forum), (2003), Mimeograph, NGO Forum, Lalmatia, Dhaka.

42. Persson, T H, (2003), "Demand for Water and Sanitation in Bangladesh", Department of Economics, Lund University, Sweden.

43. Rahman, M M, Kunal P, Chowdhury, U K, Sengupta, M K, Lodh, D, Basu G K, Chanda, C R, Roy S, Das R, Quamruzzaman, Q, Chakraborti, D, (2003), Groundwater Arsenic Contamination and Human Suffering in Bangladesh and West *Bengal, India, Strategic Management of Environmental and Socio-economic Issues.* A handbook: Guizhou Science and Technology Publishing House, Guiyang, China, 102-111.

44. Rahman, M H, Mamtaz, R and Ferdausi, S A, (1997), "Pilot solar desalination plants in Bangladesh", Water and Sanitation for All: Partnerships and Innovations, 23rd WEDC Conference Durban, South Africa.

45. Rammelt, C, (2003), *Strategies for Implementation of Drinking Water Supplies in Bangladesh,* Faculty of Technology, Policy and Management, Delft University of Technology, Delft.

46. Rammelt, C, (2005), *Personal Communication,* Faculty of Technology, Policy and Management, Delft University of Technology, Delft.

47. Shamim, I and Salauddin, K (1994), "Energy and Water Crisis in Rural Bangladesh—Linkage with Women's Work and Time", Dhaka: Women for Women.

48. Shees, K M, (2000), "Access to Safe Water", *The Independent Weekend Magazine* (Bangladesh Newspaper), April 21.

49. Smith, M M H, Hore, T, Chakraborty, P, Chakraborty, D K, Savarimuthu, X and Smith, A H, (2003), "A Dugwell Program to Provide Arsenic Safe Water in West Bengal, India: Preliminary Results", *Journal of Environmental Science and Health,* 138 pp. 289-299.

50. SOS-Arsenic.net, (2003), Scientific Studies on Dug Wells: An Easy Alternative to Obtain Arsenic Free Water, at *http://www.sos-arsenic.net/english/dugwell/dugwell3.html*

51. SOS-Arsenic.net, (2004), Arsenic mitigation update — 2004. *http://www.sos-arsenic.net/english/mitigation/updat e_miti.html*

52. SOS-Arsenic.net, (2004), Toxic effects of arsenic. *http://www.sos-arsenic.net/english/toxic_effect/*

53. Sutherland, D, Kabir, M O and Chowdhury, N A, (2001), "Rapid Assessment of Technologies for Arsenic Removal at the Household Level", *Technologies for Arsenic Removal from Drinking Water,* Workshop on Technologies for Arsenic Removal from Drinking Water organized by Bangladesh University of Engineering and Technology (BUET), Dhaka, Bangladesh and The United Nations University (UNU), Tokyo, Japan.

54. Talukder, S A (1999), "Studies of Analytical Chemistry in Drinking Water Quality and Arsenic Calamity in Groundwater in Bangladesh", In: Groundwater studies for Arsenic contamination, Phase 1: Rapid investment phase, Final report, S1, Review of existing data, British Geological Survey and Mot Macdonald Ltd.

55. Tsushima, S (2001), "Arsenic Contamination in Ground Water in Bangladesh: An Overview", at *www.kfunigraz.ac.at/wiwww/aan/news12/contamin.hml*

56. Van Geen, A, Ahsan, H, Horneman, A, Dhar, R K, Zheng, Y, Hussain, I, Ahmed, K M, Gelman, A, Stute M, Simpson H J, Wallace S, Small C, Parvez F, Slavkovich, V, LoIacono, N J, Becker, M, Cheng, Z, Momotaj, H, Shahnewaz, M, Seddique, A A, and Graziano, J H (2002), "Promotion of Well-switching to Mitigate the Current Arsenic Crisis in Bangladesh", *Bulletin of the World Health Organization,* 80 (9), pp. 732-737.

57. World Health Organization (WHO), (2001), *Fact Sheet No 210,* at *www.who.in/ inffs/en/fact201.html*

# 10

SPEECH

# US-International Climate Change Approach: A Clean Technology Solution*

*The US Administration's international engagement on climate change issues centers on five key ideas, all of which extend from and build on its own experience here in the United States. A successful international response to climate change requires developing country participation, which includes both near-term efforts to slow the growth in emissions and longer-term efforts to build capacity for future cooperation actions, absent the participation of all major emitters, including developing countries, the goal of stabilizing GHG concentrations will remain elusive. US has initiated and is participating in a range of new technology initiatives designed to meet climate and clean development goals. The article briefly highlights a few of the most significant partnerships like Group on Earth Observations, International Energy Research and Development Partnerships, The Methane to Markets Partnership and World Summit on Sustainable Development Partnerships. The article also highlights the Asia-Pacific Partnership for Clean Development and Climate and Energy Policy Act of*

* This is the speech given by Paula Dobriansky.

***2005. Meeting the challenge of the expected future growth in global energy demand and reducing greenhouse gas emissions will require a transformation in the way the world produces and consumes energy over the next generation and beyond. It will require new ways of collaborating with US partners to break through long-standing stalemates. This is why US is leading global efforts to develop and deploy breakthrough technologies for both the developed and developing world.***

Mr. Chairman, Members of the Subcommittee, thank you for the opportunity to appear before you today to address the "US-International Climate Change Approach: A Clean Technology Solution." During this afternoon's testimony, my colleagues and I will describe the numerous activities that the Bush Administration is taking to support the multiple goals of improving energy security, promoting economic growth and development, reducing air pollution, mitigating greenhouse gases and eradicating poverty.

I am particularly pleased to discuss here today Title XVI, Subtitle B of the Energy Policy Act of 2005 – which is in keeping with the Administration's practical, technology-based focus to this issue – and to outline our new Asia Pacific Partnership on Clean Development and Climate.

In his June 2001 and February 2002 climate change policy speeches, President Bush highlighted the importance of international cooperation in developing an effective and efficient response to the complex and long-term challenge of climate change.[1]

The Administration's international engagement on climate change issues center on five key ideas, all of which extend from and build on our own experience here in the United States. First, a successful international response to climate change requires developing country participation, which includes both near-term efforts to slow the growth in emissions and longer-term efforts to build capacity for future cooperation actions. Absent from the participation of all major emitters, including developing countries, the goal of stabilizing GHG concentrations will remain elusive.

Second, we will make more progress on this issue over time if we recognize that climate change goals fall within a broader development agenda – one that promotes economic growth, reduces poverty, provides access to modern sanitation and clean water, enhances agricultural productivity, provides energy security, reduces pollution, *and* mitigates greenhouse gas emissions. Countries do not look at individual development goals in a vacuum, and approaches that effectively integrate both near and longer-term goals will yield more benefits over time.

Third, technology is the glue that can bind these development objectives together. By promoting the development and deployment of cleaner and more efficient technologies, we can meet a range of diverse development and climate objectives simultaneously.

Fourth, we need to pursue our international efforts in a spirit of collaboration, not coercion, and with a true sense of partnership. This is especially true in our relations with developing countries, which have an imperative to grow their economies and provide for the welfare of their citizens. Experience has shown these countries to be quite skeptical of climate mitigation approaches that they think will divert them from these fundamental goals. It is also true that many of the largest greenhouse gas emitters are also among our most significant trading partners. They have rapidly advancing – in many cases, world class – industries and considerable technical wherewithal. We view countries like China and India as responsible partners in our efforts.

Finally, we need to engage the private sector to be successful. While the right kind of government-to-government collaboration can pave the way for great progress, we will need to harness the ingenuity, resources and vision of the private sector in developing and deploying technology.

We are putting these ideas into practice. Since 2001, we have established a range of partnerships that will address key aspects of the climate challenge while also advancing other important international objectives. We have established bilateral climate partnerships with 15 countries and regional organizations that, together with us, comprise some eighty percent of global greenhouse gas emissions. These partnerships serve as the umbrella for over 400 collaborative activities undertaken by US agencies and their partners on science, technology and policy issues. Through these partnerships, US experts are working with Australia and New Zealand to strengthen our capacity to monitor climate in the Pacific; with India to promote local level pollution and energy solutions that will have greenhouse gas intensity benefits; with Brazil to promote effective application of renewable energy; with Japan and Korea to promote greater integration of climate and energy strategies throughout Asia; and with China to enhance technical capacity for climate-related decision-making.

In addition to our bilateral partnerships, we have initiated to participate in a range of new technology initiatives designed to meet climate and clean development goals. Let me briefly highlight a few of the most significant partnerships.

## Group on Earth Observations[2]

On July 31, 2003, the United States hosted 33 nations—including many developing nations—at the inaugural Earth Observation Summit (EOS), out of which came a commitment to establish an intergovernmental, comprehensive, coordinated, and sustained Earth observation system. The climate applications of the data collected by the system include the use of the data to create better climate models, to improve our knowledge of the behavior of carbon dioxide and aerosols in the atmosphere, and to develop strategies for carbon sequestration. The United States was instrumental in drafting a ten-year implementation plan for a Global Earth Observation System of Systems, which was approved by 55 nations and the European Commission at the 3rd EOS summit in Brussels in February 2005. The United States also released its contribution through the Strategic Plan for the US Integrated Earth Observing System in April 2005.[3] The plan will help coordinate a wide range of environmental monitoring platforms, resources, and networks.

## International Energy Research and Development Partnerships

*The Generation IV Nuclear partnership,*[4] *the Carbon Sequestration Leadership Forum*[5], *the International Partnership for the Hydrogen Economy*[6], *and ITER*[7]. In the last four years, the Administration has engaged in four partnerships that lend new international emphasis to strategic technologies that can make a large contribution to our efforts to reduce greenhouse gas intensity and diversify the global energy portfolio. The State Department is working closely with DOE to engage our partners, and all of these partnerships include key developing countries as full partners in our efforts to advance these important technologies – an important capacity building function that will also serve to promote the growth of global markets.

## The Methane to Markets Partnership[8]

This partnership, launched in November of last year, focuses on advancing cost-effective, near-term methane recovery and use as a clean energy source to enhance economic growth, promote energy security, improve the environment, and reduce greenhouse gases. At the recent session, the partnership welcomed its seventeenth member, Ecuador, and now represents over 60 percent of global methane emissions. This Partnership includes an extensive project network comprised of 190 private sector, governmental and non-governmental organizations. Methane to Markets currently targets four major methane sources: landfills, underground coal mines, and natural gas and oil systems, and animal waste management. By 2015, the Partnership has the potential to deliver annual reductions in methane emissions of up to 50 million metric tons of carbon equivalent or recovery of 500 billion cubic feet of natural gas.

## World Summit on Sustainable Development Partnerships[9]

The United States has been at the forefront of efforts to move multilateral bodies toward a practical, results-focused actions centered around partnerships among governments, businesses and other organizations. Among over 20 US-initiated partnerships launched at the 2002 World Summit on Sustainable Development (WSSD) held in Johannesburg, South Africa, the United States established a "Clean Energy Initiative." The Initiative consists of four market-oriented, performance-based partnerships, including:

- **Global Village Energy Partnership (GVEP)[10]**, an international partnership with over 700 public and private sector partners with a leading role for the US Agency for International Development;
- **Partnership for Clean Indoor Air[11]**, led by the Environmental Protection Agency, addressing the increased environmental health risk faced by more than 2 billion people in the developing world who burn traditional biomass fuels indoors for cooking and heating;

- **Partnership for Clean Fuels and Vehicles**[12], led by the Environmental Protection Agency, which will help to reduce air pollution in developing countries by promoting the elimination of lead in gasoline and encouraging the adoption of cleaner vehicle technologies; and
- **Efficient Energy for Sustainable Development** (EESD)[13], led by the Department of Energy, which aims to improve the productivity and efficiency of energy systems, while reducing pollution and waste, saving money and improving reliability through less energy intensive products, more energy efficient processes and production modernization.

The United States is actively involved in other international technology development and deployment partnerships as well, including the Renewable Energy and Energy Efficiency Partnership, a WSSD partnership initiated by the United Kingdom. As the world's largest producer and consumer of renewable energy, and with more renewable energy generation capacity than Germany, Denmark, Sweden, France, Italy, and the United Kingdom combined, the United States is one of 17 partner countries in REEEP.

The United States continues to participate in the UN Framework Convention on Climate Change. The Conference of the Parties (COP) to the United Nations Framework Convention on Climate Change will hold its 11th Session in Montréal from November 28 to December 9, 2005. I will head the US delegation to this meeting. As the Kyoto Protocol entered into force on February 16 of this year, the Montréal meeting will also be the first "meeting of the Parties" (MOP) under that instrument, to which the United States will be an observer. We will continue to highlight the importance of collaborative partnerships developing and deploying technologies to meet the long-term challenge of climate change.

I am very pleased that a technology-focused approach that puts climate change in the context of broader development goals is finding favor in many parts of the world. In July, at the Group of Eight Leaders meeting, President Bush and his counterparts agreed to a Plan of Action on Climate Change, Clean Energy and Sustainable Development[14]. The Plan is based on over fifty specific, practical activities – mostly focused on technology development – that put climate change goals in the context of other development imperatives. I had the opportunity to attend a follow-up Ministerial Dialogue on November 1 that included not only Group of Eight ministers, but also ministers from eleven other key developing and developed countries. I was struck both by the strong participation from ministries responsible for energy – something that has sometimes been lacking in climate discussions – and by the very practical nature of our discussions in this setting.

## Asia-Pacific Partnership for Clean Development and Climate[15]

In keeping with the concept of practical actions and multiple benefit approaches, I would like to turn now to the Asia-Pacific Partnership for Clean Development and Climate. The Partnership is our most recent effort to promote greenhouse gas intensity reduction and other clean development goals, and we are quite excited about its potential. Deputy Secretary of State Zoellick announced plans to create the Asia-Pacific Partnership for Clean Development and Climate in July 2005. The Partnership will build on and deepen the already strong relationships we have with our five Partners: Australia, China, India, Japan, and the Republic of Korea. The six countries that currently make up this Partnership represent about half of the world's economy, population, and greenhouse gas emissions – which gives us a tremendous opportunity to find practical approaches to address these issues with our partners in a focused setting. We intend to use this opportunity to ensure that the Partnership delivers real and significant results in energy security, clean development, and greenhouse gas intensity reduction.

The Partnership's vision statement has identified a broad range of near and long-term technologies and practices that are designed to improve energy security, reduce pollution and address the long-term challenge of climate change. The Partnership will focus on voluntary practical measures to create new investment opportunities, build local capacity, and remove barriers to the introduction of clean, more efficient technologies. It is critically important to build on mutual interests and provide incentives to tackle global challenges such as climate change effectively.

We are united with our partners in recognizing that the ingenuity and energy of the private sector is crucial to our success in addressing these issues over time. This effort cannot succeed without strong private sector involvement. Working closely with the Department of Commerce and other agencies with export-oriented functions, we are actively discussing ways of ensuring that the private sector is engaged in a meaningful way in the Partnership at every stage of its work. We expect the Ministerial launch to have strong participation from the private sector.

## Energy Policy Act of 2005

The Administration welcomes the Hagel-Pryor amendment to the Energy Policy Act of 2005, which we believe will lend considerable focus and force our efforts to address climate change. This legislation is fully in line with the Administration's view that reducing greenhouse gas intensity is the best metric for measuring progress in climate change policy. In 2002, President Bush committed the United States to a comprehensive and innovative program of reducing greenhouse gas intensity by 18% by 2012. It is estimated that meeting this commitment will prevent the emission of more than 500 million tons of carbon equivalent greenhouse gases.

The approach embedded in this legislation is that the answer to the long-term challenge of climate change lies in promoting, rather than impeding, economic growth, which can, in turn, fuel the kinds of technology innovations and capital stock turnovers needed to deploy cleaner, and more efficient technologies. The legislation identifies the need to work with the major developing nations to promote cleaner technologies as they continue to work to deliver modern energy services to their people.

We are now actively working to fulfill the initial requirements of Title XVI, Subtitle B—Climate Change Technology Deployment in Developing Countries. The bill requests that we identify the major emitters of greenhouse gases, and provide a range of baseline information to the Congress on progress on greenhouse gas intensity reduction projects, obstacles to implementation, and opportunities for greater advancement. We expect to have this report to you in February, and we will use it as a basis for developing our strategy on these issues. We have active collaboration already in many of these countries, but in many cases that collaboration can be considerably strengthened. In implementing the Act, we want to ensure that our cooperation with countries is based on the practical, collaborative approach that we have developed with our partners to date. I look forward to working with your staff and that of other interested committees to ensure that we are taking a sensible and robust approach as we move forward with implementation.

The bill also requires the Secretary of State to establish an interagency working group, chaired by the Secretary, to coordinate activities under the Subtitle. We have been actively working with other agencies to ensure that this work is undertaken in a manner that complements that of other Administration efforts, including that of the National Security Council's Policy Coordinating Committee and the Trade Promotion Coordinating Committee. In addition, we expect to work closely with our colleagues at the Department of Energy and other agencies in their efforts to fulfill Subtitle A of this title.

We expect the Asia-Pacific Partnership to be one of the key means through which we implement our actions under the Energy Policy Act. In his statement at the signing ceremony for the Act, President Bush highlighted the Partnership as an innovative program that is authorized by the Act. The initiative targets the kind of fast-growing, middle-income industrializing countries on which the Act asks us to focus. China, India and Korea rank first, second, and third respectively among fast-growing industrializing economies in 2003 carbondioxide emissions – the latest data available – from the consumption and flaring of fossil fuels. In fact, depending on the data set used, these three countries alone account for roughly half of the greenhouse gas emissions among the 25 countries that we will focus on in implementing the Title. The Partnership explicitly references greenhouse gas intensity reduction among its clean development goals. We also see that a range of existing programs can contribute to these efforts, and that we can strengthen these programs and develop new strategies for achieving the objectives of the Title.

## Concluding Remarks

Mr. Chairman and Members of the Committee, I hope that my testimony this afternoon conveys the extent to which the United States is working with our partners to reduce greenhouse gas intensity, promote energy efficient technologies and advance climate science, while also placing primary importance on supporting economic growth and prosperity.

Meeting the challenge of the expected future growth in global energy demand and reducing greenhouse gas emissions will require a transformation in the way the world produces and consumes energy over the next generation and beyond. It will require new ways of collaborating with our partners to break through long-standing stalemates. This is why we are leading global efforts to develop and deploy breakthrough technologies for both the developed and developing world.

I thank you for this opportunity to testify before this Committee. I look forward to responding to any questions you may have.

*(Paula Dobriansky, Under Secretary State for Democracy and Global Affairs, Before the Senate Subcommittee on International Economic Policy, Export and Trade Promotion, Committee on Foreign Relations, Washington, DC).*

## References

[1] *http://www.whitehouse.gov/news/releases/2001/06/20010611-2.html and http://www.whitehouse.gov/news/releases/2002/02/20020214-5.html.*

[2] *http://earthobservations.org/.*

[3] *http://iwgeo.ssc.nasa.gov/docs/EOCStrategic_Plan.pdf.*

[4] *http://www.nei.org/index.asp?catnum=3&catid=1215*

[5] *http://www.cslforum.org/*

[6] *http://www.iphe.net/*

[7] *http://www.iter.org/*

[8] *http://www.epa.gov/methanetomarkets/ and http://www.methanetomarkets.org/.* Founding Methane to Markets member governments include the United States, Argentina, Australia, Brazil, China, Colombia, India, Italy, Japan, Mexico, Nigeria, Russian Federation, Ukraine, and the United Kingdom. The Republic of Korea became the 15th member in June, 2005 Canada the 16th member in July 2005, and Ecuador the 17th member in November 2005.

[9] *http://www.sdp.gov/sdp/initiative/cei/28304.htm.*

[10] *http://www.sdp.gov/sdp/initiative/cei/44949.htm.*

[11] *http://www.sdp.gov/sdp/initiative/cei/29808.htm and http://www.pciaonline.org/.*

[12] *http://www.sdp.gov/sdp/initiative/cei/29809.htm and http://www.unep.org/pcfv/main/main.htm.*

[13] *http://www.sdp.gov/sdp/initiative/cei/28304.htm.*

[14] *http://usinfo.state.gov/ei/img/assets/4756/PostG8_Gleneagles_Communique.pdf*

[15] *http://www.state.gov/s/d/rem/50326.htm.*

# Section III

# Corporate Responsibilities & Environmental Accounting

# 11

# Corporate Social Responsibility

*Paul Watchman and Malcolm Forster*

***The current global concern is to protect environment or sustainable development. At the same time, industrial development is important for economic growth and employment generation for providing a better quality of life. Therefore, both development of the corporate sector and protection of environment go hand in hand. However, rapid development of company and industrial activities, without proper precautionary measures for protection of environment, are causing pollution. Hence, it has become a necessity for a company to comply with the regulatory norms for prevention and control of pollution to the extent that the regulatory norms are to be complied with by adopting clean technologies and improved management practices. Commitment and voluntary initiatives of industry for taking care of the environment will help in building a partnership for pollution control. Today's globalization raises new challenges regarding how to develop trading relationships that support the UN's Millennium goals of reducing poverty and ensuring greater environmental security. Companies are aware that they can contribute to sustainable development by managing their operations in such a way as***

*Source: The Icfai Journal of Environmental Law, July 2005.* 

***to enhance economic growth and increase competitiveness; ensuring environmental protection; and promoting social rights. Thus, Corporate Social Responsibility (CSR) is a concept whereby companies voluntarily integrate and include social and environmental measures in their business activities. Corporate Social Responsibility is one of the responses to the imbalances resulting from the acceleration of the globalization process. The company aims to protect and restore the environment by minimizing the use of resources and energy, by decreasing waste and harmful emissions, from its activities.***

## Introduction

Corporate Social Responsibility (CSR), an understanding of the social, ethical and environmental impact of the activities of a business, is becoming an increasingly important preoccupation of business managers. Often thought of primarily as a response to moral or ethical imperatives, many of the most significant social obligations are imposed on business as a direct result of the application of environmental and public international law principles. CSR is part of an international drive towards transparency and accountability of business activities and a way of monitoring how businesses perform against environmental, ethical and social indices.

At present, both the UK government and the EU see CSR as voluntary, an aspect of good corporate governance, although there is pressure from some quarters for a mandatory system. Whether or not CSR remains voluntary, business must face up to the challenge that it offers.

There is nothing particularly threatening about CSR, but companies must take it seriously and get it right. The information which companies are collecting as part of CSR is being devoured by pension fund managers, Non-Governmental Organizations (NGOs), stakeholders, media and government.

CSR need be neither a hypocrite's charter nor a stone to weigh down corporate enterprise. It can be a competitive tool or a dirty laundry list of legal liabilities. The choice is yours.

## What is Corporate Social Responsibility?

The European Commission describes CSR as:

'... a concept whereby companies integrate social and environmental concerns in their business operations and in their interaction with their stakeholders on a voluntary basis.'[1] Similarly, the UK Department of Trade and Industry (DTI) regards CSR as involving:

'... businesses looking at how to improve their social, environmental and local economic impact, their influence on society, social cohesion, environmental and human rights, fair trade and on the ways that fairness can be corrupted. CSR is an issue for large multinationals and for small, locally based businesses.'

## Why Bother about Social Responsibility?

Social responsibility on the part of businesses is not a new phenomenon. Since, the 19th century, there have been many shining examples of business investing in model housing, healthcare and education for workers and their families. The modern vision of CSR, however, is a more systematic and integrated program to assess the environmental and social impact of the business and to manage it in a strategic manner.

Globalization of business has heightened the necessity for companies to improve their awareness of the effects of their increasingly complex and diverse operations in a wide variety of geographical regions, some of them unfamiliar to management until recently. The growth in market liberalization in many parts of the world has given rise to international operations in companies (some of which are relatively modest in size), which would never have ventured beyond their domestic markets before.

There are a number of factors which have increased the importance of a company's ability to make a satisfactory account of its activities in the field of social responsibility.

The most immediate incentive lies in the steadily increasing tempo of regulatory supervision. This is perhaps most keenly felt in the environmental and health and

---

1 Promoting a European Framework for Corporate Social Responsibility COM (2001) 366.

safety fields, but stringent requirements are also imposed in relation to money laundering, financial dealings of suspected terrorists etc. Many of these requirements impose upon businesses positive duties of documentation and information, together with the institution of monitoring and reporting systems.

On a wider spectrum, the impulse for CSR comes from stakeholder pressure. This takes many forms, of which the most common are:

- Pressure from shareholders who may regard a sound CSR performance as an indicator of good corporate governance and an essential element in an attractive investment profile for the company;
- Pressure from ethical investment vehicles, which now routinely survey a wide range of CSR issues in assessing a company's performance. Ethical investment has soared in the past five years with investment in ethically screened UK funds standing at some £4bn in 2003 compared with £1.5bn in 1997[2];
- Ethical issues have become an important factor in lending decisions by both public and private sector financial institutions. For many years, the World Bank, other multilateral development banks and export credit agencies have been insisting that borrowers respect environmental and human rights values. Recently, the private banking sector has responded by subscribing to the Equator Principles, which replicate many of these concerns;
- Pressure from customers and potential trade partners who are themselves increasingly coming under pressure to demonstrate that their commercial interactions underpin CSR objectives;
- Potential for much wider exposure of the company's activities through the rapid dissemination of material over the internet; and
- Pressure for increased company reporting on general CSR issues on a voluntary basis, developing from the now firmly established trend for reporting on environmental performance.

---

[2] Source: World Business Council for Sustainable Development.

A further impetus may come from financial services regulators. Recent studies have suggested that some of the largest companies in the world were not making adequate provision for the costs which might be required to adjust their carbon emissions to meet climate change concerns. Nor were they alone in this miscalculation, as similar studies of the attitudes of investors revealed that the financial risks threatened by climate change issues had not been taken into account by investors, who remained either ignorant or complacent about the long-term implications for the value of their investments in this respect. Such relaxed attitudes may not endure much longer. Pressure is growing for increased transparency and disclosure about the environmental and other social effects of company activities, both from investors themselves and from rating agencies anxious to respond to this new investor interest.

An equally significant driver is the expectations of civil society. Governmental promotion of CSR, while usually falling short of mandatory requirements, is on the increase. Even more pertinent is the activity of non-governmental organisations (NGOs), many of which have secured a very high degree of visibility and respectability for their published surveys and reports. Among these are Amnesty International, FTSE4Good, the Dow Jones Sustainability Index, etc.

This activity is replicated in a number of business sector initiatives, such as Business in the Community's corporate responsibility index and the Association of British Insurers' guidelines on corporate social responsibility.

A more remote, but not inconceivable, incentive lies in the possibility that a company which can arguably be held not to have complied adequately with some CSR norms at least may find itself the target of civil action in plaintiff-friendly jurisdictions.

## Governmental and International Developments in the Field of CSR

## OECD Guidelines for Multinational Enterprises

The Organization for Economic Cooperation and Development (OECD) adopted these guidelines in September 2000.

They commit businesses to contributing to economic, social and environmental progress with a view to achieving sustainable development and respect for human rights. They are also called upon to engage in capacity-building in the countries in which they operate and not seek exemptions from local environmental, health, safety, labor or tax laws, even when these are offered to them, nor should they engage in any 'improper involvement' in local political activities. Businesses should also ensure that timely, regular, reliable and relevant information is disclosed as to their activities, structure, financial situation and performance, using high-quality standards for disclosure, accounting and audit. This information should include the social, ethical and environmental policies of the enterprise and other codes of conduct to which the company subscribes.

Companies should seek to avoid compulsory or child labor, discrimination in all its forms and respect the right of their employees to be represented by trade unions and other bonafide representatives of employees, and engage in constructive negotiations, observing standards of employment and industrial relations not less favorable than those observed by comparable employers in the host country.

In relation to the environment, businesses should take due account of the need to protect the environment, public health and safety and generally conduct their activities in a manner contributing to the wider goal of sustainable development. They should also assess, and address in decision-making, the foreseeable environmental, health, and safety-related impacts associated with the processes, goods and services of the enterprise over their full life cycle. Where these proposed activities may have significant environmental, health, or safety impacts, and where they are subject to a decision of a competent authority, the company should prepare an appropriate environmental impact assessment.

As to corrupt practices, businesses should not, directly or indirectly, offer, promise, give, or demand a bribe or other undue advantage to obtain or retain business or other improper advantage.

## The European Union

The European Commission has set out a number of basic principles for the EU's program relating to CSR. These are that:

- CSR is to remain an essentially voluntary exercise;
- CSR practices must be transparent and credible;

- The Community's own strategy in CSR matters should be focused in the areas where its involvement can add value;
- This strategy should adopt a balanced and comprehensive approach to CSR, including not only consumer concerns, but also economic, social and environmental issues;
- The strategy should pay particular attention to assisting Small and Medium Sized Enterprises (SMEs); and
- The strategy should support other international legal obligations (such as environmental treaties, basic human rights and labor conventions, etc).

The Commission sees its strategy as:

- Seeking to improve awareness of the beneficial impact of CSR on business and society at large;
- Developing a mechanism for exchange of information about good CSR practice, both between businesses and across member states;
- Developing CSR management skills;
- Assisting SMEs to develop CSR practices suitable to their particular capacities and needs;
- Developing codes of conduct in relation to corrupt practices, environmental, human rights and labor issues;
- Promoting the development of management standards for CSR, modelled perhaps on the Eco-Management and Audit Scheme (EMAS) environmental management standards;
- Promoting 'triple bottom line' (economic, social and environmental) reporting by companies, especially those employing more than 500 people (as foreshadowed in the Commission's 2001 communication on the EU strategy for sustainable development); and
- Exploring the possibility of introducing a labelling system for products or enterprises which reflect CSR values.

The Commission has set up and chairs an EU multi-stakeholder forum on CSR, consisting of representatives of employers, trade unions, civil society, and trade and professional associations. It will try to promote transparency and harmonization of CSR practices by exchange of information, coordinating existing initiatives within the EU and trying to form a common EU position on which to base discussions with third countries, while also identifying areas where further action is needed at EU level.

Similarly, the Commission proposes the integration of CSR values into all EU policies, including those on economic and social affairs, enterprise, the environment, consumer protection, public procurement, external affairs, development policy and trade.

As a result, in December 2002, the Council called on member states to:

- promote CSR at national level in parallel with the development of a strategy at community level, in particular by making companies aware of its benefits and highlighting the potential results from constructive cooperation between governments, business and other sectors of society;
- Continue to promote the dialogue with social partners and civil dialogue;
- Promote transparency of CSR practices and tools;
- Exchange information and experiences regarding their policies;
- Integrate CSR into national policies; and
- Integrate, where appropriate, CSR principles into their own management.

## The UK Government

As part of its commitment to reform company law, the UK government published (in July 2001) a White Paper entitled *Modernizing Company Law*[3]. This document addressed the role of directors in a modern context, which is seen as combining the traditional obligation to serve the best financial interests of the shareholders with a broader awareness that the company is likely to perform most effectively when a wider ranger of economic, environmental and social issues and influences are taken into account.

[3] Cmnd.5553 (2002).

A further proposal would oblige larger companies to publish annually an Operating and Financial Review (OFR). This duty would be placed upon:

- PLCs with a turnover of more than £50 mn or a balance sheet showing net income and reserves of more than £25 mn; or
- Limited companies with a turnover of more than £500 mn or a balance sheet showing net income and reserves of more than £250 mn.

The OFR would discuss and analyze the business and the main trends and influences which shape its performance and which are likely to impact upon that performance in the future. The obligation to disclose an issue arises if failure to do so in a clear and unambiguous manner would, in the honest opinion of the directors, be expected:

> *'to influence members' assessment of the company and hence the decisions they may take, either directly, or indirectly as a result of significance that the information has for other stakeholders and thus the company.'*

In this way, it is thought, shareholders, investors and others will be able to form a clearer picture of the likely or possible impact of ethical and environmental issues on the profitability of the company and of any possible threats to its operations, e.g., by reason of its investment in countries which are likely to be subjected to economic sanctions or of its involvement in activities which might spark a boycott of its products, ethically-driven disposal of its shares or (as in the case of some companies targeted by animal welfare groups) the withdrawal of banking, insurance or other services by suppliers anxious to avoid 'guilt by association'.

The White Paper envisages that, in order to compile the OFR, directors will be required to have in place a system that enables them to gather and assess a wide range of information. According to the draft Companies bill annexed to the White Paper, the OFR must include a statement of:

- The company's business activities and its objectives, together with a statement of its strategy;
- A review of the company's business over the year in review and an assessment of its position at the year-end, including any material changes or events; and

- Any known event or trend which may significantly affect its performance or investment program in the future.

In addition, the directors are to include other information when this appears to them to be necessary to achieve the objective of the OFR (to enable members of the company to make an informed assessment of its operations, financial position and its future business objectives and strategies). Into this class falls information about governance issues, relationships with business partners and questions relating to matters which might have an impact on the reputation of the company (including environmental, social and ethical questions). In particular, the directors should consider including information on the company's policies relating to:

- Employment by the company;
- Environmental issues relevant to its business; and
- social and community issues relevant to its business.

The White Paper regards it as the directors' business to ensure that all this information is available to them and to provide mechanisms to provide it where these are currently lacking. It also envisages that directors may wish to assure themselves of the accuracy of the information, for example, by having it verified by third parties.

An Operating and Financial Review Working Group has been established to help provide guidance on how directors can assess whether items will be required to be included in the OFR. The Working Group issued a consultation document in the summer of 2003. Since then, the government has announced its intention to publish draft regulations on the requirement for an OFR under its existing powers. The draft regulations are anticipated to be issued in summer 2004.

Government departments are also taking part in a number of initiatives aimed at developing a methodology for preparing CSR policies and reports. The DTI, the Department for Environment, Food and Rural Affairs (DEFRA) and the Department for International Development (DFID), together with the Environment Agency, are represented on the steering group of the SIGMA program, a joint project by the British Standards Institute, Accountability and Forum for the Future which has produced a set of guidelines, combining a set of guiding principles to help commercial organizations understand the relevance of sustainability to their activities and how they can respond, supplemented by a management framework which provides a model for integrating CSR issues into corporate decision-making.

## The CORE Campaign

CORE (the Corporate Responsibility Coalition) sponsored a private member's bill in the last session of parliament aimed at introducing mandatory requirements for companies to act responsibly on humanitarian and environmental issues. The bill (partly sponsored by Friends of the Earth and Amnesty International) expresses the frustration felt in some parts of the NGO community with the lack of government progress in the CSR field and, in particular, with its support for a voluntary approach. The bill would apply to all companies (except those with an annual turnover of less than £5m). It would impose a duty on companies to conduct their activities in accordance with local laws in each jurisdiction in which it operates (not necessarily a high standard in itself) but also in accordance with 'international agreements, responsibilities and standards' relating to:

- The environment;
- Public health and safety;
- Sustainable development;
- Employment;
- Human rights; and
- Consumer protection.

The bill would make reporting mandatory in respect of the social, environmental and economic impacts of its activities in the year under review and of its plans for the future. It would also provide for parent company liability for:

- Serious physical or mental injury; or
- Serious harm to the environment caused by the company's activities, where these are conducted in a manner inconsistent with applicable standards so as to fail to ensure the safety of the claimants or the environment.

Attempts to reintroduce the bill (renamed the Performance of Companies and Government Departments (Reporting) Bill) have proved unsuccessful so far. An Early Day Motion has been tabled in which CORE calls on the government to act quickly to introduce laws requiring companies to report on their social and environmental performance as well as creating a duty of care on company directors.

## What are the Sources of CSR Obligations in International Law?

Obligations relevant to CSR can be found in:

- Major environmental treaties, such as:
  - The Convention on Biological Diversity;
  - The Climate Change Convention and the Kyoto Protocol; and
  - Flagship wildlife treaties, such as the Migratory Species Convention, the Wetlands Convention, etc.
- The principal human rights treaties, such as:
  - Universal Declaration on Human Rights;
  - International Covenants on Economic, Social and Cultural Rights and on Civil and Political Rights (ICESCR and ICCPR);
  - UN Conventions on Racial and Sexual Discrimination;
  - UN Convention on the Rights of the Child; and
  - The 'core conventions' of the International Labor Organization.

In addition, the UN Sub-Commission on the Promotion and Protection of Human Rights has recently completed work on: *Norms on the responsibilities of transnational corporations and other business enterprises with regard to human rights.* The Commission on Human Rights considered these for the first time in April 2004. The norms are largely exhortatory, in that most of the provisions merely urge transnational corporations to respect the obligations already set out in the national and international human rights law, including those set out above. There are, however, a number of significant features which are worthy of note.

- The human rights protected by the norms extend beyond the widely ratified texts referred to above and include the Optional Protocols to the ICCPR (which provide for individual petition to the Commission and abolition of the death penalty). They also include the still-controversial 'right to development'.
- The obligation to ensure equal opportunity contains a reference to 'affirmative action' measures to overcome discrimination in the past.
- The obligation to provide a living wage requires consideration of the worker's need for adequate living conditions with a view towards progressive improvement.
- The obligation to respect national laws is extended to certain policy instruments, including 'the public interest' (undefined) and developmental objectives.
- There is an obligation to ensure that goods and services provided are not used to subvert human rights.
- There is an ill-defined duty not to deal in materials which are harmful or potentially harmful to consumers.
- The environmental obligations extend beyond the present international law requirements, in that they require observance of the precautionary principle, 'bioethics' and compliance with the 'wider goal of sustainable development'.
- The norms contain an imprecise obligation on states to set up legal and administrative structures to ensure that the norms are implemented.
- The norms envisage 'reparation' being made to those whose human rights have been adversely affected by transnational corporations.

## How can these CSR Obligations be Relevant in Day-To-Day Business?

### Environmental Issues

Accounting for environmental performance in a transparent manner is at the core of CSR. It is, however, true that the environmental impacts of some industries, such as the energy and extractive industries, are more amenable to measurement than those of other sectors and, for this reason, have a longer history of calibrating environmental performance than have less environmentally intrusive industries. Nevertheless, while it is important to bear in mind the general principles of environmental law and the form in which they have been transposed and applied into legal norms in different jurisdictions, this is less easy than it might appear, because some of these principles are somewhat inchoate and are still in the process of development, and they frequently interact in unexpected ways.

The areas of environmental concern that should be the subject of consideration when a CSR policy is being designed or a report compiled are set out below.

### Sustainable Development

Sustainable development is less a legal principle (if only because its content is understood differently by each environmental constituency) than an overarching objective of environmental law. The 1992 UN Conference on the Environment and Development (the Earth Summit) gave rise to the most commonly used definition of the concept of sustainable development, namely:

> *'Development which meets the needs of the present without compromising the ability of future generations to meet their own needs.'*

Sustainable development is, therefore, about ensuring a better quality of life for all members of society now and for future generations to come and involves stakeholders at all levels, from national governments to public and private limited companies. Effective protection of the environment is just one of the objectives of sustainable development, the others being social progress, the prudent use of natural resources and maintenance of high, and stable, levels of economic growth and employment.

The adoption of the concept of CSR is seen as positively contributing to achieving sustainable development.

### Legacy Issues—The Polluter Pays Principle

It is a long established principle of environmental liability that the 'polluter pays' to put right any harm, or detriment it causes to the environment.

Increasingly, legislation looks to impose liability for harm caused regardless of how long ago it occurred. Prime examples of this are: The introduction of the UK contaminated land regime, which imposes primary liability for the cleanup of contaminated land on the person who caused[4] or knowingly permitted[5] the presence of the contamination where they can be traced; and the recently adopted environmental liability directive, which establishes a framework of environmental liability to prevent and remedy environmental and biodiversity damage based entirely on this principle.

In the absence of similar legislation in the jurisdictions in which the company operates, a company will need to consider whether there is firstly the potential for such issues to exist and, if so, its obligations to report and how it intends to manage the issues arising.

## Climate Change Issues (Including Emissions Trading, Clean Development Mechanisms, etc)

Recognition of the impact of greenhouse gas emissions on climate change, and particularly the environment, has been high on the political agenda for sometime. The UN Framework Convention on Climate Change (UNFCCC) 1992 was opened for signature at the Earth Summit, with the objective of:

> *'Stabilization of greenhouse gas concentrations in the atmosphere at a level that would prevent dangerous anthrogenic interference with the climate change.'*

All parties to the UNFCCC are subject to general commitments to respond to climate change. These general commitments were extended by the 1997 Kyoto Protocol to the UNFCCC, which resulted in the industrial nations committing themselves to legally binding targets and timetables for the reduction of GreenHouse Gas (GHG) emissions. The protocol permits the use of flexible mechanisms to achieve reductions in GHG emissions, including emissions

---

[4] Causing requires involvement in an active operation or chain of operations that results in the presence, or continued presence, of the contaminant and can encompass a failure to act.

[5] Knowingly permitting requires both knowledge of the presence of the contaminant and the power to prevent it being there. The definition is potentially wide enough to draw in subsequent owners and occupiers where they become aware of the presence of the contaminant.

trading[6], joint implementation[7] and clean development mechanisms[8] by the Annex I parties to meet their emissions targets.

These international developments are also driving national and supranational initiatives that impact upon companies. In July 2003, the EU adopted a directive establishing a mandatory Emissions Trading Scheme (EUETS), which will initially set caps on carbondioxide emissions from more than 10,000 industrial installations within the EU and the 10 accession countries. Additional installations, and a wider range of GHG, are anticipated to be drawn into the second phase of the scheme, which will run 2008 to 2012. On a national level, the climate change levy has been introduced as one of HM government's initiatives to reduce GHG emissions. This is essentially a tax on businesses' use of energy generated from non-renewable sources, and is designed to encourage more efficient use of energy.

Aside from requirements to comply with legislation controlling GHG emissions, companies may feel the impact of climate change as a result of falling profits in sectors affected by severe weather events (particularly the tourism and insurance sectors). They will also be affected by related policy and legislative developments in the product liability and energy areas and the results of calls by shareholders to improve social and environmental performance.

## Biodiversity

The recognition of the need to conserve the biodiversity of life on earth, its sustainable use and the fair and equitable sharing of benefits arising from the use of genetic resources was also the subject of the Earth Summit and resulted in the signing of the UN Convention on Biological Diversity.

---

6 An emissions trading scheme involves the allocation of allowances to participants to emit GHG. Participants are given the flexibility to trade allowances between them, so that if the participants emit more than their allowances, they can purchase allowances from other participants in the market. In this way, reductions are achieved in the most cost effective way possible.

7 Joint implementation allows the industrialized nations to implement projects that reduce emissions in other industrialized nations. The resulting emissions reduction can then be used by the investing party to help meet its target and a corresponding subtraction is made from the host's emissions target.

8 The clean development mechanism allows the industrialized nations to implement projects that reduce emissions in less industrialized nations. As with joint implementation projects, the resulting emissions reduction can then be used by the investing party to help meet its target. It also helps the host nation achieve sustainable development and contributes to the objectives of the UNFCCC.

The convention is underscored by a number of related treaties, and relies heavily on action by national governments. In the UK, HM government has called on companies owning or managing land to contribute to conservation of biodiversity through the development of a biodiversity action plan. Advice issued by Earthwatch indicates biodiversity action plans should involve, as a minimum, the conducting of an ecological site survey; the preparation of an action plan with appropriate targets and deadlines for implementation of measures to conserve, and where possible, enhance biodiversity on the company's sites; implementation of the plan; and monitoring and reviewing and reporting of performance.

## Compliance with Environmental Laws

This is one of the fundamental objectives of CSR reporting, involving as it does the identification and management of risks. Environmental laws encompass a wide range of issues including the condition of water, air and land and also the conservation of flora and fauna and protection of human health. By identifying and tracking the relevant environmental laws that apply to a company's operations, the company is in a position to ensure compliance and so avoid committing environmental offences and corresponding harm to the company's reputation and image.

The recent line of case law involving foreign claimants seeking to argue that UK-based parent companies should be liable for losses and damages suffered by them as a result of the acts and defaults of their foreign subsidiary companies also serves to illustrate that companies need to be as alive as to what is happening overseas in their name as to that which is happening on their own doorstep.

## Waste Management (Minimization, Reuse, Recycling and Deposit Safely)

The need to reduce the amount of waste being sent to end disposal is a key priority in EU environmental policy because of the environmental harm techniques such as landfilling and incineration can cause. Waste policy and legislation, therefore reflects the need:

- To reduce the amount of waste generated by improving manufacturing methods and influencing consumers to demand greener products and less packaging;

- For increased recycling and reuse where waste cannot be prevented, through the use of producer responsibility measures and the use of waste taxation; and
- For improved final disposal and monitoring by increased environmental regulation of waste incinerators and landfill sites.

Waste minimization and reuse is also recognized as being a key issue for any CSR policy and companies should consider their ability to reduce the amount of their waste arisings e.g., by encouraging electronic rather than paper documentation and filing methods, segregation of waste streams and taking steps to ensure waste is recycled and reused wherever possible.

## Energy Efficiency

Given its alliance with both climate change and sustainable development initiatives, the need to achieve increased energy efficiency is also high on the political agenda. Further, saving energy saves money as significant improvements in energy efficiency can usually be achieved for low capital outlay and with a short payback period. It also helps to make industry more competitive.

In the UK, aside from the introduction of the climate change levy in April 2001, HM government indicated in its energy White Paper *Our energy future – creating a low carbon economy* that pursuing energy efficiency initiatives is one of the ways in which it envisages delivering its target of cutting the UK's carbondioxide emissions by 60% by 2050, with real progress by 2020. Indeed, it anticipates achieving half of the extra carbon savings needed by 2020 through energy efficiency. As part of this, an energy efficiency implementation plan is due to be published in April 2004 setting out how the strategy will be delivered. Aside from the climate change levy and the introduction of the UK emissions trading scheme, HM government envisages changing building regulations to ensure energy efficiency measures are included in the design of new buildings, encouraging the work of the Carbon Trust[9], and has introduced the enhanced capital allowances scheme, which enables businesses to claim 100% first year capital allowances on investments in energy saving technology.

[9] The Carbon Trust was launched in April 2001 and is developing and implementing programs to accelerate the take up of energy efficiency in the non-domestic center.

## Stewardship of the Environment

The concept of stewardship of the environment is about promotion of environmental sustainability, and in particular ensuring that the impact of a company's activities on the environment are considered and managed so as to ensure that the minimum possible harm is caused to the environment and that the concept of intergenerational equity is respected.

The following examples illustrate how environmental issues in the context of CSR may affect companies.

- A European home products company discovers it is supplying goods made from tropical timber which is not grown from sustainable forests, but is wild-logged from tropical rain forests in a manner which is destructive of the reproductive capacity of the forest and which has the effect of substantially reducing the habitat available to endangered species.
- A company's local subsidiary in a southeast Asian country discharges effluent to surface water in compliance with a local operating permit, which contains fewer conditions and higher effluent limits than those applied to exactly similar plant in western European country. These discharges are completely in compliance with local law. The discharges, however, result in fish mortality and other biodiversity damage and there is evidence of increased incidence of gastric illness among local people using the river downstream for the purposes of washing, drinking water and irrigation.
- An internal business review proposes that production from the company's German plant (which is a major emitter of carbondioxide) should be transferred to a plant which the company has just bought outside Moscow, on the grounds that labor costs will be lower and that the Russian Federation has not ratified the Kyoto Protocol, so carbondioxide emission controls will not have to be installed.
- The management of an energy company visits a refinery in an African country where some basic petrochemicals are produced. In accordance with local practice, sludges and some liquids are disposed of in unlined lagoons. There is some evidence that the soil and groundwater in the immediate locality of these lagoons is being contaminated by oils and tars, which are

capable of migration through groundwater. There is no evidence of any impact on workers or local inhabitants, nor is there any impact on wildlife.

## International Law of Human Rights

CSR obligations in the international human rights treaties are often expressed in terms of broad general principle, but they are capable of very precise application and companies can easily be criticized for failing to comply with the expectations that the treaties raise.

## The Right to Freedom from Discrimination

The obligation to avoid discrimination is a very broad one, prohibiting discrimination on grounds of race, ethnic origin, nationality, religion, gender, political opinion, physical disability, etc. Many companies will have policies which deal with discrimination in some sectors, especially sexual discrimination (which is also the subject of an International Labor Organization (ILO) convention), but it is important critically to examine the whole of the company's practice to ensure that the policy is carried into effect and is reinforced by sound training programs, recruitment and promotion policies. For example, are there provisions in the company's personnel policy to meet the special needs of women in the neonatal or nursing period? If not, it could be argued that the company is failing to comply with the obligation. Also, difficulties may occur where the company has subsidiaries and business partners in countries where the local law permits or entrenches discrimination on a gender, ethnic or other basis. A corporate social policy that merely aims at compliance with local law will not protect the company from criticism from international pressure groups, for example.

Note that particular problems may arise if the workforce includes indigenous peoples, who are entitled to recognition of their special rights and cultural identity. An ILO convention contains specific rules to be observed in relation to such people. A common issue arises out of the fact that indigenous people often own rights in land or natural resources on a communal basis and their interests are not always adequately respected in cases where land is acquired by public authorities for major projects. Companies should be alert to the necessity to ensure that indigenous communities have been properly treated before engaging in such projects.

## The Right to Personal Safety and Security

Most companies will have a health and safety policy which covers in broad terms the management of the risks to which its employees and others are exposed. Nonetheless, care must be taken to identify groups at special risk, for example, pregnant women or the physically disabled. There is a potential overlap with the obligation to avoid discrimination in such cases. A frequent trap for companies in connection with this right is inadequate control of security personnel. In some countries, security personnel may routinely behave in a manner that would be regarded as tantamount to assault in other jurisdictions. A company that had failed to give clear directions as to the level of physical force that should be employed by security personnel, may find itself in breach of the obligation. There are internationally agreed instruments regulating how security personnel should behave.

## The Prohibition of Slavery and Forced or Child Labor

Slavery within the meaning of the UN convention still exists in several parts of the world and companies operating in those regions must be alert for indications that their local subsidiaries or business partners are profiting from slave labor. Debt bondage (by which the debtor agrees to work off a loan from his or her employer at ruinous rates of interest) is not uncommon in some regions and industrial sectors. Other countries use forced labor in public infrastructure projects or benefit from the labor of prisoners working in inhumane conditions. Some agricultural societies still operate on a feudal basis. Migrant workers are also frequently exploited in the labor market of the countries where they work. The ILO conventions discourage the employment of children below the compulsory school leaving age, although once again, a company should be careful about mere compliance with a local law that reduces that age below what is internationally regarded as acceptable. Employing a twelve year-old child in 'light work' may meet local laws, but is unlikely to be regarded sympathetically elsewhere.

## The Prohibition of Torture, Cruel, Inhuman or Degrading Treatment or Punishment

In some parts of the world, it is usual for employers to use corporal punishment on workers in their employ. Companies should be alert to detect such practices (and other forms of exerting physical or mental pressure on employees) in their local subsidiaries or business partners.

## The Right to Work

Does the company policy provide for protection against arbitrary or unjust dismissal (there may be an overlap here with the duty to avoid discrimination)? Is the hiring policy even-handed between differing groups or communities? Is there a guarantee of equal pay for equal work (another ILO convention requirement)? Does the company recognize trade unions and the right to strike?

## The Right to Freedom of Peaceful Assembly and Association

Widely recognized in developed countries, trade unions and labor associations are treated much less sympathetically in some other regions. The ILO conventions impose strict requirements in this context and companies should ensure that their policies respect the employee's right to organize and do not permit any form of sanction to be levied on workers who engage in union activity. In some countries, failure to announce these rights in local languages may also result in a failure to eliminate discrimination. In a few countries, the company may even find it necessary to protect the union representatives against reprisals, e.g., where union activity is unlawful or frowned upon by government. In such cases, the company may wish to develop alternative mechanisms for meeting worker concerns.

## The Right to Participation in Political Life

Does the company's policy and practice permit employees to play a part in the political life of the country, region or community? Is there reasonable latitude in practice to afford employees time to engage in public service without losing pay, status or promotion prospects?

## The Right to Rest and Leisure

Does the company's pattern of working (organization of shifts, working hours, etc) provide workers with enough rest and free time to maintain their physical and mental health? In particular, are workers given adequate holidays (including religious and other festivals observed in the region)? There is also an overlap with the obligation to protect family life.

## The Right to an Adequate Standard of Living

Does the company operate a pension scheme and other appropriate benefits for workers who have to leave work for reasons beyond their control? Does the company respond sympathetically to workers who are obliged to be absent from work because of the illness etc. of children or other dependents? It should be recalled that the family circle in respect of which such attendance is customary may be significantly wider in some countries than in the first world.

## The Right to Education

Does the company's employment policy encourage young people in the community to curtail their education e.g., by targeting young workers because they are cheaper to employ? Does the company's training program help workers to develop their innate skills and aptitudes? Is training offered on a truly non-discriminatory basis?

In December 2003, the Business Leaders Initiative on Human Rights was established by a leading group of international businesses that seeks to explore ways in which businesses can implement the principles of the Universal Declaration of Human Rights and the United Nations code of practice on business and human rights.

## Developing a Successful CSR Policy Framework

A threshold question for the company to decide is the objective of the CSR policy.

## To Comply with Local Law Only

As mentioned above, this is not always an entirely safe option in jurisdictions where local standards are lower than those which would be acceptable to investors and other stakeholders in the state where the company (or its parent) has its principal place of business. Furthermore, the OECD and EU principles on CSR regard it as axiomatic that the company should be aiming for a standard of performance which is higher than that mandatory under local law. Stress is laid on the voluntary assumption of the company of the willingness to 'go the extra mile'.

## To Apply the same Standards Worldwide Regardless of Local Minimum Requirements

Often described as the 'Bhopal solution', this approach requires that the same standards (e.g., of environmental protection or worker safety) be employed wherever the company operates, regardless of whether they may be immeasurably higher than those required or usually observed in a particular jurisdiction. This approach simplifies the management of the company's policies on CSR issues, but may not be without its dangers. For example, if worker health functions are not genuinely devolved to a local subsidiary but are effectively delivered as a global service from headquarters, any failure to match up to requirements applicable there may result in claimants 'forum-shopping' in the courts of the headquarters' state.

## To Consciously Seek to Attain Higher Standards than those Required in any Jurisdiction

CSR does not insist that such an approach be motivated only by good corporate citizenship or altruism, but recognizes that it may be an important element in promoting the profitability or efficiency of the company or even as a means of securing competitive advantage.

The company should aim to arrive at a policy that is realistically achievable, rather than aim too high and give observers the opportunity to comment on shortfalls in performance. The company may wish to use one of the international instruments as a yardstick (the Universal Declaration is often chosen in this context). In a proper case, the company should not be shy of involving responsible NGOs in the design of the policy: Some industry sectors (such as insurance and mining) have already developed CSR codes which can be used as models.

It is vitally important to obtain the endorsement of senior management, ideally with responsibilities identified at main board level, even though the most effective vehicle to delivery of CSR objectives may not be at the worldwide level, but rather by operating division or business grouping (depending on management structure). Similarly, human resources or corporate affairs might not be the most effective 'home' for CSR in a company whose business management is strictly channelled into almost autonomous business units.

The company should map out the issues or geographical regions where it is most likely to be exposed to CSR failures. Particular consideration should be given to raising awareness for local management in areas where CSR performance would require a departure from usual business behavior and providing support for (and surveillance of) responsible officers. Suppliers, subcontractors and other business partners (such as potential joint venture partners) are also potential areas of exposure in CSR terms.

All CSR programs should contain a review and remedial function to identify shortcomings and to put them right. Documenting compliance is an essential element of telling a good story on CSR – documenting failure to respond to failures revealed is the worst of all possible worlds!

## What has the Law to do with CSR?

Business activities are described, facilitated and confined by the law. There is little that a business does which is not a legal matter and as such has legal impacts.

The bedrock of a sound CSR policy is legal compliance with international minimum standards for environmental and labor protection.

It is an essential precondition to designing a workable CSR policy to understand the nature and impact of the legal obligations on your business, both in each jurisdiction in which your company operates and in respect of public international law that may be more onerous than local laws. It is no use claiming (however accurately) that the company complies with local standards, if these are nowhere near as demanding in environmental or social terms as those in western countries. In many, indeed almost all, areas of the CSR field, international law obligations set the agenda for business activities.

Understanding legal concepts, rules and process are essential to business and CSR. Limited liability, lender responsibility, shareholder tights, *forum non conveniens,* responsibility of parent companies for subsidiaries, civil liability principles (including no fault liability) all are central to understanding a company's responsibilities and liabilities.

Legal reasoning and legal rigor is an important method and discipline to bring to CSR.

## Where Lawyers can Add Value

The keystone of CSR is compliance with international and national legal standards. Advising on the applicability, interpretation and application of laws is quintessentially what lawyers are trained to do. Lawyers should bring an independent, objective and professional evaluation of:

- The legal risks faced by the company;
- Which legal obligations are binding on it;
- To what extent they are being complied with; and
- How to deal with legal problems if they arise.

In the CSR field, there are three principal areas where legal support is of particular relevance:

## Exposure Mapping

Before a company embarks on the difficult task of crafting a CSR policy, it should have an accurate chart of the often unfamiliar waters into which it is sailing. Rocks and hidden reefs lurk offshore to entrap unwary mariners. For example, management, in its anxiety to collect all the information needed to prepare an OFR, may not appreciate that, by gathering data on the environmental performance of the company, they may actually be creating or triggering legal liabilities. The criminal law offence, for example, of 'knowingly' permitting pollution is dependent upon establishing that the company has knowledge of the environmental harm or wrong in question. That knowledge may be acquired from answers to a questionnaire sent out to a business unit or subsidiary as part of the parent company's CSR exercise.

Similarly, in many jurisdictions, information about soil contamination or groundwater pollution must be disclosed to the relevant regulatory body. Again, a company may find itself guilty of a criminal offence for failing to disclose information which lies festering in the in-tray of a junior employee.

As a final example, there is a school of thought which urges that parent companies should be made responsible for damage caused to the environment or abuse of human rights by their subsidiaries. This may seem a radical step, but it is in fact a challenge which companies must now face up to in the wake of the *Cape* asbestos case in the House of Lords[10] and the US decision in *Wiwa v Royal Dutch Petroleum (Shell)*[11].

Forming an accurate impression of the scope of the company's exposure to risk in the CSR field is largely a matter of calibrating its performance and objectives against the various domestic and (particularly) international norms, which the stakeholders, from their varying perspectives, will expect to see observed. Companies should resist the temptation to narrow the focus of the review. CSR is not a field in which ignorance is bliss.

## External Validation

External advisers should validate CSR policies and the mechanisms that support them in order to check the accuracy of what is reported by the company and the robustness of the conclusions which are drawn from the data. At present, an increasing number of management accountants, environmental consultants, NGOs or academic experts are making themselves available to discharge this function. This is all very worthwhile as far as it goes. Management accountants have developed very sophisticated structures for gathering commercial information from even the remotest parts of complex corporate structures and for evaluating that information against financial and commercial norms. Environmental consultants have, over the past decade in particular, become highly literate in the commercial impact of environmental costs. NGOs, on the other hand, tend to be more sectoral in their perspective and are inclined to err somewhat on the side of pessimism about a company's performance: Some, indeed, operate within a culture which makes it difficult for them to take an objective view of corporate activities or aspirations. Academic experts, while often possessed of considerable erudition in a defined field, rarely have the broad commercial experience to orient what they discover in the context of the business environment. Also, they infrequently have access to logistic backup to enable them to deliver a quality product in a timely manner.

10 *Lube v Cape plc* [2000] 1 WLR 1545

11 [2000] US App

The lawyer cannot and should not seek to replicate the very different skill sets offered by these professionals. There are, however, limits to the extent to which the traditional audit approach that is adopted in relation to financial and other commercial disciplines can be deployed effectively in relation to CSR issues. Frequently, those companies that have been farsighted enough to make a CSR report and to have it externally verified are obliged to submit to have that verification qualified, sometimes in a manner that seems very sweeping. A qualification to the effect that: environmental and social data are subject to many more inherent limitations than financial data, given both their nature and the methods used for determining, calculating and estimating such data is, in fact, entirely accurate and reflects to the credit of the auditor, but it does little to reassure the publisher of the report or the reader.

There is, however, a particularly important role that the lawyer can play in this context. An allegation by an unfriendly stakeholder that the company is in breach of a domestic or international legal standard in a CSR matter would, if it were substantiated, be the clearest possible indicator of the company's failure to take CSR obligations seriously. Therefore, it is of the utmost importance that the allegation be critically examined in relation to the facts which underpin it in order to determine whether there has indeed been such a breach. That is an achievable objective in most cases that can do much to remove uncertainty, and it is quintessentially a legal task. As such, the person to advise on the matter is an experienced external adviser, familiar with the substance of the law concerned; with the temperature of the market and with the tenor of the public mind; who can defuse the allegation (if it should prove to be unfounded); but who can also focus on the need to tell the client things that may be unpalatable but which are ultimately in its best interests to acknowledge and address. CSR policies and reports are rightly the subject of intense scrutiny by the public and NGOs. They should be as legally watertight as possible.

## Fire-Fighting

The very scope and diversity of CSR issues means that it is almost inevitable that a thorough mapping of exposure and the operation of a well-designed and efficiently operated data collection system will reveal areas where the company is failing to achieve its CSR aspirations. In such a case, the lawyer's task is to provide advice about the company's legal obligations, the extent to which they have been transgressed, whether the breach is capable of being remedied and how this should be achieved, as well as the steps that should be taken to minimize legal liability and damage to reputation. For example, it may be possible to negotiate a remediation strategy with an environmental authority that will diminish its appetite for prosecution or a means may be found to disclose information which meets legal requirements without provoking public criticism.

*(Paul Watchman is a Member, Environment, Planning and Regulatory Group, 65 Fleet Street, London EC4Y 1HS, He can be reached at paul.watchman@freshfields.com*

*Malcolm Forster is a Joint Head of the Public International Law Group, 65 Fleet Street, London EC4Y 1HS, He can be reached at malcolm.forster@freshfields.com).*

# 12

# Extended Producer Responsibility for Environment Protection: The Concept, Relevance and Lessons

*C S Shylajan and G Radha Krishna*

***This paper discusses the concept of Extended Producer Responsibility (EPR) which has emerged as a policy tool for managing the environment by promoting recycling, and thus reducing waste. The concept implies that responsibilities, which were traditionally assigned to consumers and waste management authorities, are to be extended to the manufacturer of the products. It is based on the 'producer pays' principle. It provides incentives to producers for making Design for Environment (DfE) changes to products that would reduce waste management costs. This paper reviews important EPR policy tools implemented in countries around the world and discovers that many industrialized countries have implemented EPR policies and have been successful in achieving the targets of reducing waste. Some important principles which are relevant for implementing EPR programs effectively are also discussed.***

*Source: The Icfai Journal of Environmental Economics, November, 2006.* 

## Extended Producer Responsibility: The Concept and Goals

The concept of Extended Producer Responsibility (EPR) has emerged as an important environmental policy tool in many industrialized countries to manage environmental issues, especially for managing solid waste (OECD, 2006). The Organization for Economic Cooperation and Development (OECD) defines EPR as, "an environmental policy approach where the producers' responsibility, physical and/or financial, for a product that is extended to the post-consumer stage of a product's life cycle" (OECD, 2000). The concept of EPR implies that responsibilities, which were traditionally assigned to consumers and waste management authorities (mostly municipalities and municipal corporations), are to be extended to the manufacturer of the products. It provides incentives for producers to make Design for Environment (DfE) changes to products that would reduce waste management costs (OECD, 2006). The concept has emerged as "an environmental protection strategy to reach an environmental objective of a decreased total environmental impact from a product (packaging waste, for instance), by making the manufacturer of the product responsible for the entire life cycle of the product and especially for the take-back, recycling and final disposal of the product" (Lindhqvist, 2000). It is supposed to result in an improvement in the design of the products in favor of reducing waste, reuse or recycling, material conservation, pollution reduction and lower toxicity (Schwarz and Gattuso, 2002). The EPR program attempts to incorporate product disposal or recycling costs as an upfront fee. The fee is based on the estimated cost of recycling or recovering materials. It is based on the 'producer pays' principle. The manufacturers who cut down on packaging waste pay less fees, which ultimately leads to waste reduction.

## EPR Policy Instruments

Important EPR policy instruments which have been carried out in many developed countries are:

- Mandatory product take-back programs with recycling targets.
- Voluntary product take-back with recycling rate targets.
- Advanced Recycling Fees (ARF).

- Advanced recycling fees with a recycling subsidy.
- Deposit Refund System (DRF).
- Environmental Labeling System.
- Recycled content requirements.
- Material use restrictions.

Most mandatory EPR programs have two components, namely:

- Fee paid by product manufacturers on targeted products and
- Specific take-back goals for each targeted material or product (Schwarz and Gattuso, 2002).

EPR has been most frequently applied to packaging and electronics. There are both mandatory and voluntary EPR approaches that have been implemented in many countries. In Europe, many countries have implemented various EPR programs to fulfill their legislative requirements. For instance, the European Union has set minimum waste recovery and recycling targets by product type. In December 1994, the European Union issued a directive on packaging and packaging waste to recover waste and promote recycling. This legislation places direct responsibility, and specific packaging waste reduction targets, on all manufacturers, importers and distributors of products on the EU market. As per this directive, the European Union members are required to pass a national legislation to reduce packaging waste and also set minimum recovery and recycling targets. EU targets are revised every five years. The most recent legislation passed in 2004, established 2008 targets of 60% for overall waste recovery and 55% for recycling (OECD, 2006). To meet the requirements of this legislation, manufacturers, importers and distributors are required to either develop and implement their own take-back schemes for their product packaging or join industry-driven, non-profit organizations that are established to organize the collection, sorting and recycling of used sales packaging.[1]

Many European countries have their own mandatory EPR laws on recycling and recovery targets (Table 1). The UK, for instance, passed the Producer Responsibility Obligation (Packaging Waste) Regulation in 1997, and the Packaging

[1] www.packaging-waste.com

**Table 1: Recycling or Waste Recovery Targets under the EPR Program in Selected Countries**

| Nation | Products | Recycling or Waste Recovery Targets |
|---|---|---|
| Austria | Batteries, Refrigerators, Packaging | 20-70% |
| Finland | Packaging | 82% of packaging recovery |
| Germany | Batteries, Packaging | 60-75% recovery |
| Luxembourg | Packaging | 55% recovery |
| The Netherlands | Appliances, Batteries, Packaging | Packaging recycling: 65% Appliance/ battery recovery: 90% |
| Portugal | Packaging | 50% recovery by 2005 |
| Singapore | Packaging (Voluntary) | 80% waste reduction through recycling |
| Slovenia | Packaging | 78% by 2010 |

*Source: Schwartz and Gattuso (2002).*

(Essential Requirements) Regulations in 1998. The producers are required to recover and recycle a specific percentage of their packaging waste each year, as shown in Table 2. It can be observed that both recovery and recycling targets are increasing every year, especially since 2004. The UK regulations divide the producer responsibility into four categories: (1) manufacturer, (2) converter, (3) packer/filler,

**Table 2: UK Packaging Recovery and Recycling Targets**

| Year | Total Waste Recovery Targets (%) | Recycling Targets (%) |
|---|---|---|
| 1998 | 38 | 7 |
| 1999 | 43 | 11 |
| 2000 | 45 | 13 |
| 2001 | 56 | 18 |
| 2002 | 59 | 19 |
| 2003 | 59 | 19 |
| 2004 | 63 | 59 |
| 2005 | 65 | 61 |
| 2006 | 67 | 63 |
| 2007 | 69 | 66 |
| 2008 | 70 | 66.5 |

*Source: OECD (2006).*

and (4) seller. According to this regulation, a manufacturer is a person who manufactures raw materials for packaging and is responsible for 6% of the recycling obligations. A converter is a person who uses or modifies packaging materials in the production or formation of packaging. He is responsible for 9% of the recycling requirements. The packer/filler is a person who puts goods into packaging, and he is responsible for 37% of the recycling requirements. Lastly, the seller, who supplies packaging to a user or a consumer of that packaging, is responsible for 48% of recycling obligations.

## EPR Policy: Best Practices

Many developed countries have adopted and implemented various types and combinations of EPR policy instruments to reduce the environmental impacts of products on environments. Japan, Korea and Taiwan have introduced EPR programs for containers and packaging, among other items. Germany, the Netherlands, Sweden, Denmark, Finland and many OECD countries have adopted EPR programs for a variety of packaging products, for safe disposal. It is observed that for most of the countries, the success rate has been very high. The Green Dot program is one of the popular EPR schemes in the world to take-back packaging waste. The Green Dot is a trademark used on packaging. It is established by manufacturers, retailers and importers as a non-profit system to fulfill government take-back requirements. It was first developed in Germany, in 1991. It is currently being implemented in 19 European countries and Canada. Germany, Norway, Sweden, UK, Belgium, Luxembourg, France, Poland, Austria, Hungary, Spain, Czech, and Slovenia are some of the countries that have implemented the Green Dot program. Participation in the Green Dot program allows manufacturers, who sell their products in the EU, to comply with the EU's Packaging Directive of 1994, which requires producers to recover their packaging waste. In case a manufacturer does not participate in the program, the company is required to individually arrange for the take-back of its packaging waste. It is estimated that over 200 million people dispose their packaging via a collection system set up by the Green Dot organization.[2]

Many countries also have voluntary take-back schemes implemented by manufacturers to reduce the cost of waste management. Some examples of the products are electronics, plastics, single-use camera, laser print cartridges, paints,

---

[2] *www.packaging-waste.com*

solvents, beverage containers, pharmaceuticals, lubricants and pesticides, etc. Dell Computers' take-back program, Hewlett Packard's toner cartridge return program, or Nike's 'Reuse-a-Shoe' program are some of the voluntary programs set up by companies. Sony Display Device in Singapore has set up a voluntary program to reduce solid waste. Through the three common methods of waste minimization—reduce, reuse and recycle, Sony's efforts to minimize waste have contributed to an annual saving of about S$1 mn (Exhibit 1).

**Exhibit 1: Voluntary EPR by Sony Display Device, Singapore**

Since 1990, Sony Display Device, Singapore, has been manufacturing Cathode Ray Tubes (CRT) for television. CRT manufacturing is a complex operation that involves the use of a large number of raw materials, chemicals, and components. CRT manufacturing generates a lot of solid waste, such as sludge, waste glass and various packaging materials. The three basic waste minimization strategies of reduce, reuse and recycle, are extensively employed at every step of the manufacturing process. Sony decided to reuse the packaging material several times before their functional properties declined to the extent that disposal becomes inevitable. Arrangements are made with customers to return the polyfoam material and wooden pallets back to Sony for reuse. Sony saves S$1 mn annually from this operation. Sony also has a very efficient system for the recycling of glass waste. The largest component of the waste generated in CRT manufacturing is glass. This accounts for almost 50% of the total waste generated in the CRT manufacturing process. Sony implemented a system whereby the waste glass is collected, and is then sent to a glass vendor in Japan for recycling. Collection bins are provided at the production areas to facilitate segregation at the place of waste generation. The continuous monitoring and training has resulted in glass recycling rate of 97% every year through the three common methods of waste minimization—reduce, reuse and recycling, Sony's efforts to minimize waste sent to the landfill have contributed to an annual saving of about S$1 mn.

*Source: www. nea.gov.sg/cms/rcd/guidebook/sony.pdf*

Deposit Refund System (DRS) is another important EPR tool which has been practiced in many countries for safe disposal of solid waste. It can be defined as a 'refundable product charge'. A deposit is imposed at the time of purchase of a product and allows the consumer for a refund of the charge at the end of the product cycle.

The purpose of DRS is that the consumer must return the product or its container for waste recovery, recycling or for safe disposal. Many European countries have introduced deposit-refund system for electric bulbs, beverages and PET (Polyethylene Terephthalate) bottles, lead and accumulators, Nickel-Cadmium batteries, car hulks, soft drinks, beer, wine, spirit containers, glass, aluminium cans, pesticides or fertilizer containers, etc. They have been successful in implementing the deposit-refund system due to: (1) the high awareness of consumer regarding the environmental effects of the products (2) proper legislative mechanism and its implementation as part of their waste management strategy. Table 3 gives the details of the high recovery rate for a gamut of products from beer and PET bottles with 99% recovery and minimum of 60% recovery for electric bulbs.

**Table 3: Success Rate of Deposit-Refund System in Some European Union Countries**

| Country | Product | Deposit Rate (in Euros) | Return of Products for Safe Disposal (percentage observed in the system) |
|---|---|---|---|
| Austria | • Electric bulbs | 0.7267/bulb | 60% |
| Denmark | • Plastic bottles | 0.2907/packaging | |
| | • Beer and PET plastic bottles: | | |
| | – 0.5 liters | 0.3353/bottle | 99% |
| | – 1.5 liters | 0.6035/bottle | 99% |
| | • Nickel-Cadmium batteries | 16.0939/kg battery | 75% |
| Finland | • Soft drinks, beer, wine, spirit containers: | | |
| | – Bottle cases | 2.3543/case | Very high |
| | – Glass bottles | 0.0775-0.3875/bottle | Very high |
| | – Metal cans | 0.1682/case | Very high |
| | – Plastic bottles | 0.0775-0.3875/bottle | Very high |
| The Netherlands | • Bottles for beer, soft drinks, milk and dairy products | | |
| | – Soft drinks (PET) 1-2 liters | 0.4538/bottle | 95-99% |
| Sweden | • Glass, PET bottles, aluminum cans | | |
| | – Aluminum cans | 0.0547/can | 90% |
| | – PET bottles | 0.4372/bottle | |

*Source: www1.oecd.org/scripts/env/ecoinst/index.htm*

Successful deposit-refund schemes operate in the OECD countries, i.e., Austria, Belgium, Denmark, Finland, France, Germany, the Netherlands, Norway, Sweden, Turkey and the United States. Studies from USA also show that the deposit-refund system is more cost-efficient compared to other policy instruments such as recycling, subsidies as well as recycled content standards (Palmer and Walls, 1997; Palmer *et al.* 1997).

Apart from the deposit-refund system and product take-back system, there are many other incentives that have been in practice in many countries. The main types are environmental labeling system and product fees. In the case of environmental or eco-labeling scheme, businesses voluntarily label their products to inform consumers and to promote products determined to be more environment-friendly than other functionally and competitively similar products. The result is the development of products with lesser environmental impacts, and thus reduced waste, pollution, and waste quantities. Germany introduced its eco-label program called Blue Angel in 1977, making it the first country to implement a national ecolabeling scheme (Chakrabarti, 2003). Many countries like Canada, Japan, Thailand, South Korea and Taiwan have introduced eco-labeling programs to promote recycling, and reduce pollution from waste generation (see Table 4). The Government of India introduced an eco-labeling scheme known as 'Eco-mark' in 1991, for easy identification of environment-friendly products. This scheme helps consumers in prompting manufacturers to adopt clean and eco-friendly technologies and environmentally sound disposal of used products. It provides incentives for manufacturers and importers to reduce adverse environmental impacts of their products. So far, 16 consumer product categories have been awarded the 'Eco-mark' label in India[3]. However, the success of an eco-labeling scheme depends on many factors such as, consumers' environmental awareness, producers' commitment for environment protection, cost-effectiveness, bureaucratic procedures for getting license, etc. For eco-labeling to be an effective tool, the label must have substantial market penetration in order to affect a significant number of producers. The procedures for obtaining a license to use eco-label should also be simplified. Another important factor in

[3] The product categories are soaps and detergents, paper, food items, lubricating oils, packaging materials, architectural paints and power coatings, batteries, electrical/electronic goods, food additives, wood substitutes, cosmetics, aerosol propellants, plastic products, textiles, fire extinguishers and leather (www.envfor.nic.in/ cpcb/cpcb.html).

**Table 4: Overview of Eco-Labeling Progams**

| S. No. | Eco-label | Started in | No. of Product Categories | No. of Products Awarded | Methodology |
|---|---|---|---|---|---|
| 1. | Germany: BLUE ANGEL | 1977 | 88 | 3,355 | Modified LCA* |
| 2. | Canada: ECOLOGO | 1988 | 50 | 3,000 | LCA |
| 3. | Japan: ECOMARK | 1989 | 68 | 4,647 | LCA |
| 4. | The Nordic Council: NORDIC SWAN | 1989 | 46 | 1,200 | LCA |
| 5. | India: ECOMARK | 1991 | 16 | 1 | LCA |
| 6. | Thailand: GREEN LABEL | 1993 | 31 | 187 | LCA |
| 7. | South Korea: ENVIRONMENTAL LABELING | 1992 | 79 | 319 | Modified LCA |
| 8. | Taiwan: GREEN MARK | 1992 | 72 | 1,058 | LCA |

Note: * Life Cycle Assessment is the assessment of the environmental impact of a given product or service throughout its lifespan.

*Source: Chakrabarti (2003).*

the success of any environmental labeling scheme is its ability to cover its costs and therefore, stay in business (Chakrabarti, 2003).

Many countries have direct product charges, which are special fees for handling products with difficult disposal requirements or adverse environmental impacts. Advanced Recycling Fees (ARF) is an advance disposal fee. It is a tax, whose assessment is based on product sales, and used to cover the cost of recycling. Examples are charges on batteries, packaging, tyres, libricant oils, scrapped cars, fee on imported aluminum cans, etc. Generally, the revenue generated through these product charges are used for financing collection, treatment, recovery and disposal of problem products, such as batteries, tyres, etc. Countries like Hungary, Latvia, Bulgaria, Lithuania, Korea, Japan, China, Thailand and many OECD countries have introduced product charges for safe disposal of problem products.

## EPR: Lessons

The EPR concept has been successful in many countries, which have introduced mandatory take-back and recycling requirements. The Netherlands and Korea, for instance, have successfully introduced the Waste Electronics and Electrical Equipment (WEEE) Program, which mandated that retailers take-back old electronic and electrical goods for meeting recycling obligations. A study conducted by OECD in February 2006 shows that all of the EPR programs they have studied, reported high recovery and recycling rates (OECD, 2006). However, it shows that policies that directly target design for environment are likely to be very costly and difficult to implement or enforce. The above study cautions that "policymakers need to keep in mind that multiple policy instruments are necessary for efficiently accomplishing multiple environmental goals." At this juncture, the following OECD's guiding principles have more relevance for implementing an effective EPR program by any government (OECD, 2000).

- EPR policies and programs should be designed to provide producers with incentives to incorporate changes upstream, at the design phase, in order to be more environmentally sound.
- Policies should stimulate innovation by focusing more on results than on the means of achieving them, thus allowing producers a flexibility with regard to implementation.
- Policies should take a life cycle approach and be directed at producing life cycle benefits, even if they focus on the post consumer phase, so that environmental impacts are not increased or transferred somewhere else in the product chain.
- Responsibilities should be well-defined.
- Policies should be product-specific. One type of program or measure is not applicable to all products, product categories or waste streams.
- A communication strategy should be devised to inform all the actors in the product chain, as well as consumers, about the program and to enlist their support and cooperation.
- To enhance a program's acceptability and effectiveness, consultation of stakeholders about its goals and objectives, as well as estimates of its costs and benefits, should be conducted.
- Local governments should be consulted in order to clarify their role and obtain their advice concerning the programs' operation.

- Both voluntary and mandatory approaches should be considered, with the aim of meeting national environmental goals and objectives in the best way possible. The process of developing and implementing an EPR strategy, and putting it into operation, should be based on transparency.

## Conclusion

The above discussion shows that the concept of EPR has emerged as an important policy tool for managing environmental issues, particularly, waste management. EPR policies are incentive mechanisms for manufacturers for taking adequate measures to reduce, recover and recycle waste. Both voluntary and mandatory approaches have been implemented in many countries with the aim of meeting national environmental goals. As firms compete to add value for their customers, many of the companies voluntarily implement EPR policies such as, product take-back and environmental labeling programs. Voluntary EPR policies can provide a competitive advantage to a company in many ways. The experience of many companies has demonstrated that they can improve resource efficiency and cut down waste handling and disposal costs. Companies like Sony and HP have enhanced their relationship with customers through green initiatives such as end-of-life voluntary take-back schemes. Since consumers have become more environment conscious than before, the companies can enhance their reputation by going green, and thus gain a winning edge over their competitors. EPR policies can be implemented in other countries, especially developing countries, if: (1) the government and companies collaborate together to educate and create the consumer awareness regarding the environmental impacts of products and (2) the government provides legislations which can be implemented (3) companies become good corporate citizens by creating and developing green products or eco-friendly products.

*(C S Shylajan is a Faculty Member at the Icfai Business School (IBS), Hyderabad, Andhra Pradesh, India. He can be reached at shylajan@ibsindia.org and*

*G Radha Krishna is a Faculty Member at the Icfai Business School (IBS), Hyderabad, Andhra Pradesh, India. He can be reached at grkrishna31@rediffmail.com)*

## References

1. Chakrabarti (2003), "Comparative Ecolabeling Schemes of Selected Countries", Occasional Paper 2, Center for Development and Environment Policy, Indian Institute of Management, Calcutta.

2. Kinnaman T C (ed.) (2003), *The Economics of Residential Solid Waste Management*, Ashgate: USA.

3. Lindhqvist T (2000), "Extended Producer Responsibility in Cleaner Production: Policy Principle to Promote Environmental Improvements of Product Systems", IIIEE Dissertation 2000: 2 (Lund: International Institute of Industrial Environmental Economics, Lund University), www.iiiee.lu.se/Publications.nsf/DisplayDissertation?OpenPage

4. National Environment Agency, *Guidebook on Waste Minimization for Industries* (Draft), www.nea.gov.sg/cms/rcd/guidebook/sony.pdf

5. Organization for Economic Cooperation and Development (2000), *Extended Producer Responsibility: A Guidance Manual for Governments*, OECD: France.

6. Organization for Economic Cooperation and Development (2006), *EPR Policies and Product Design: Economic Theory and Selected Case Studies*, OECD: France.

7. Palmer K, Sigman H and Walls M (1997), "The Cost of Reducing Municipal Solid Waste" in Kinnaman, T C (ed.) (2003), *The Economics of Residential Solid Waste Management*, Ashgate: USA.

8. Palmer K and Walls M (1997), "Optimal Policies for Solid Waste Disposal Taxes, Subsidies, and Standards" in Kinnaman, T C (ed.) (2003), *The Economics of Residential Solid Waste Management*, Ashgate: USA.

9. Schwartz Joel and Dana Joel Gattuso (2002), "Extended Producer Responsibility: Reexamining its Role in Environmental Process", Policy Study 293, Reasons Foundation, Los Angeles.

10. *www.oecd.org/scripts/env/ecoinst/index.htm*

11. *www.envfor.nic.in/cpcb/cpcb.html*

# 13

# Measuring Sustainability and Green Reporting: The Role of Environmental Audit

*Arup Choudhuri*

***Caring for the environment has assumed an important place in the world culture. People from all walks of life, including the corporate citizens, have become aware of the impacts that pollution of air, ground and water can have on the quality of life, the enjoyment of the earth's beauty, and the physical well-being of its inhabitants. In view of this, governments all over the globe have entered the picture with laws and regulations that are intended to correct and cure the effects of past violations of good environmental practices, and which are designed to prevent future violations of good environmental disciplines. Accounting, being a social discipline, cannot thrive in a vacuum. It has been enriched with various environmental guidelines and standards in different corners of the globe. By using environmental accounting as a strategic weapon, companies can prosper by acquiring brand equity and goodwill. To keep an efficient environmental accounting system in place, the two major areas to be considered are valuation vis-à-vis quantification of environmental issues and***

*Source: Accounting World, July 2005.* 

***control through proper environmental auditing. This article brings out the major environmental auditing issues and proper valuation of environmental perspectives in order to achieve corporate social responsibility.***

It is apparent that the area of environmental auditing is probably one of the most dynamic and important subjects to come to the attention of the corporate world. It is an area with which internal and external auditors, and engineers as well as accountants must become familiar to make their companies survive in the contemporary world of competition. It is an enormous area under the influence of companies, corporate stakeholders, environmental protection organizations and departments of law and justice at all levels of government in all the nations, be it underdeveloped, developing or developed.

In both accounting and auditing, quantification and valuations become a primary issue. In measuring the cost of environmental degradation and associated liabilities, some objective and realistic basis is needed. In the US, both the Environmental Protection Agency (EPA) and the Securities and Exchange Commission (SEC) have strict requirements for disclosure. Though the third world governments do not enforce many regulations as far as environmental laws and reporting are concerned, the urge to introduce environmental accounting and auditing standards is being felt in most of these countries. As a result of official responses to both the Brundtland Commission (World Conference on Environment and Development—WCED), in 1987 and the United Nations Conference on Environment and Development in 1992 (the 'Earth Summit'), most governments have adopted sustainable development as a national goal. An emerging debate is about how economic sectors or businesses can contribute to this objective. This has resulted in a number of concepts such as 'corporate sustainability' and 'corporate environmental responsibility' as well as a plethora of proposals to monitor the progress being made towards corporate sustainability.

Quantitative valuation, accounting and disclosure of environmental issues by a company is regarded a strategic move to outweigh the competitors today.

As the 20th century comes to an end, many businesses have made serious efforts to change their products and processes and became responsible 'green' companies. Often, however, these changes have been tough to accomplish. Attitudes have often proved as rigid as manufacturing practices.

In assessing current managerial attitudes toward a greener business, perhaps the greatest change to be noted is the new spirit in managing ecological concerns in developed nations. Seldom do managers of big corporations routinely damn these newer public needs and regulations as simply added demands on a company's time, energy, and money; the newer perspective, increasingly adopted by companies, both large and small, is to incorporate environmental factors into standard production and service activities in such a way as to create a win-win situation with customers and other environmental stakeholders. In short, companies are finding that good corporate citizenship sells.

This transition has not occurred overnight. In most firms, the changes in attitude and behavior have been gradual and often arduous to implement. In earlier years, such changes often comprised only minimal shifts in product design and small adjustments in operating practices. However, technological considerations sometimes unexpectedly raised the ante; new technologies introduced to meet new regulations might necessitate significant changes in operating practices, along with concurrent major investments in monetary terms. These types of decision situations have led to greater managerial involvement and helped shift companies into more advanced stages of responsiveness to the green issues.

In search of whether there is a general pattern to the way companies have responded to environmental concerns, Warren B Brown and Necmi Karagozoglu (1998) looked back over the last quarter of a century and identified four broad stages of corporate responses to the laws and regulations that gave them concrete expression.

The first was an early period of resistance to pollution controls. Companies would legally challenge the appropriateness of new environmental legislation or, at best, comply grudgingly. The costs of compliance were played up as a major concern, and many firms talked of job losses if the new regulations were enforced. Correspondingly, actual corporate adjustments at this stage tended to be the 'end-of-pipe' variety, which meant doing only enough to ensure that the final products met the new requirements in not polluting the air, water, or soil.

Companies also complained often and loudly that environmentalists were vastly overstating the problems, or that technical solutions were not yet available. With such attitudes, it is not surprising that few firms then invested time and money to analyze their total operations from an environmental viewpoint, or to make significant redesigns in their products and operations to eliminate the sources of the problems. Most of the corporations in India and other underdeveloped nations are in this stage or at its threshold.

The second stage of corporate response reflected a major shift in management attitudes. Companies began to speak of the importance of environmental stewardship. Managers began to shift gears and recognize the deep-seated reality that environmental concerns had become a significant force in business environments. They also saw that through regulations and laws, these concerns could become a serious economic and technical determinant of the way their firms operated. Even more fundamentally, many managers themselves had begun to realize the validity of various ecological concerns and wanted their companies to be viewed as responsible corporate citizens in dealing with them. In short, they wanted to take pride in their companies' products, services and operations.

The begrudging compliance of the earlier stage gave way to a more positive attitude, one in which companies publicly expressed understanding and acceptance of the social and environmental dimensions that accompanied their economic role in society. 'Environmental stewardship' became a fashionable phrase. Skeptics said that many of the new corporate pronouncements were simply window dressing or cheap rhetoric. Nevertheless, optimists noted that many corporate managers were taking an active part in their companies' environmental affairs and in business seminars on the subject. The focus of operations was being moved toward the prevention of pollution, not just its control. Moreover, the stewardship concept itself kept evolving and broadening; for example, accounting systems were used to help understand all the costs of a company's environmental impacts and the specific life cycle costs of its products, rather than only direct manufacturing costs.

The stewardship concept opened the door for many companies to move beyond the first two stages. In the third stage, a company becomes regularly active in this area. It moves beyond adaptive responses, however encompassing, and strives continuously to leverage environmental issues and use them as strategic weapons. Not only is it a good corporate citizen in the earlier sense of cooperating with public stakeholders and complying with regulations, but it has also begun to move ahead of its industry and its consumers—to create win-win situations in which it profits by its ecologically oriented strategies while taking pride in being a good environmental steward. Product redesign is often one path used to achieve these dual ends. Besides an appealing new look and perhaps some new features, a redesigned product may also use new materials that are recyclable, can be produced with less waste through a more efficient manufacturing process, and may involve less use of toxic chemicals for cleaning purposes.

A well-known example in this area is from America's hamburger master, McDonald's, which changed its burger wrapping from plastic to paper in response to environmental pressures and reaped a rich harvest of good publicity in return. Less well-known was the broader context of the move; the wrapping change was actually only one of several interrelated efforts McDonald's made in instituting a far more extensive, though less visible, waste-reduction strategy. It is also at this stage that many companies, like McDonald's, seek partnerships with regulatory and other environmental groups, easing their transitions to new requirements, technologies, and investments.

The fourth stage of corporate environmental response is one that few companies have reached as yet. This stage characterizes a focus throughout all activities on managerial practices for a sustainable environment. A broader and somewhat idealized perspective is involved here, one that demands more total corporate involvement than any of the earlier stages. Companies act as agents of change in both the economy and society, working to develop business practices that are indefinitely sustainable from an environmental standpoint as well as being economically sound.

Many social and ecological dimensions of this stage are not traditional in business. For example, these newer sustainable development concepts may strive to integrate such factors as soil erosion, population growth, the decomposability of materials, and the overexploitation of renewable resources. However, despite its apparent breadth, sustainable development is not just a pie-in-the-sky theory.

Executives throughout the industrialized world have begun to recognize both the validity of the underlying concerns—particularly in terms of long-run, 'next-generation' issues, and real opportunities for revenue growth through creative action.

It is also becoming clear to businesses that a growing number of 'green consumers' worldwide actively support the concepts inherent in an environmentally responsible management and a sustainable development philosophy. They tend to focus their purchases on products from firms that share their environmental concerns. In brief, they put their money where the 'green'. is.

They would also like to see much more progress in this area, such as greater numbers of companies in the US and other developed nations of Europe, South East Asia and Japan are moving more rapidly into Stage 4 activities. With the passage of years, simple pollution control is no longer enough. Environmental issues have shifted, and critical new orientations for the business are environmental stewardship and sustainable economic development. This seems especially true for consumers and managers with a global outlook. The growing demands for rapid economic development in emerging nations and the limitations of the earth's resources give an added impetus to incorporating these modern philosophies in the modern firm.

Although many managers are not eager to take on environmental issues that reflect global problems, it is actually the corporate world that is being looked to for leadership. In the eyes of many, businesses are the only organizations that have the material resources, technology, global reach, and managerial skills to pull it all together. Couple this reality with a widespread distrust of politicians and government bureaucracies, and we can see why business should become the vehicle that leads our economies into Stage 4 activities. The hoped-for scenario is that modern corporations will provide both motivation and leadership for the new era of sustainable economic development.

The framework given above is, of course, only a brief outline, but there are several indications that such environmentally focused management practices are taking root. The questions it raises are:

- How far have we actually progressed?
- Are many win-win strategies really being developed?
- Are appropriate investment in environmental activities being made?
- Are the senior managers of competitive companies really changing their attitudes?

These questions cannot be answered right now without making an extensive survey, but it is high time we addressed environmental issues, and strategies may be formulated to create a win-win situation.

Today's accountant assumes the role of the queen bee in the organizational setup. He is at the hub of the Accounting Information System (AIS), a subset of the Management Information System (MIS). The AIS measures, processes and communicates financial information about an identifiable economic entity to various stakeholders inside and outside the company. In the post World War II economy, the discipline of accounting has undergone a metamorphosis and has assumed a social role; it is not only about the recording and processing of transaction and calculation of profit at the end of the year, accounting is about valuations of all the tangible and intangible resources used for a business in addition to being a major decision tool for all levels of management. Environmental accounting and auditing is a relatively new entrant in the domain of accounting. It is now argued that in the future, managing environmental risk and investment opportunities effectively will make the difference between companies' outperforming their competitors and lagging behind, or disappearing altogether (Kiernan and Levinson, 1997). Eco-industrial revolution has given a new shape to the discipline of accounting and 'environmental accounting' and 'green reporting' have become two major areas of research in both the corporate as well as the academic world. In the realm of environmental accounting, as discussed earlier, two major issues are valuation and control.

As far as valuation is concerned, some theoreticians laud full cost accounting, i.e., accounting for a corporate entity's internal and external costs generated as a result of its economic activity (Canadian Institute of Chartered Accountants, CICA, 1997). Specifically, these theoreticians focus on the generation of external environmental costs such as pollution and, further, link this to the existing debate regarding sustainable development. Indeed, the whole issue of evaluating external costs associated with environmental change within corporate accounts is judged to be underdeveloped in a recent survey by De Koning-Martens and Van der Ende (1997). Hence, it may be said that full cost accounting can be extended and developed and an indicator of corporate sustainability that emerges from this process can be proposed. As an empirical illustration, an example for air pollution and the UK corporate sector may be provided, although, in principle, the approach can be extended to other forms of pollution. Sustainable development is undoubtedly a complex notion that is open to numerous interpretations, and many issues surrounding the usefulness and reliability of full cost accounting remain unresolved. It may be argued that, far from being impossibly confusing, it is possible to take practical steps towards measuring corporate sustainability through full-cost valuations.

Controversy arises when we consider how sustainability is to be achieved, i.e., what are the conditions for attaining sustainable development. While there is no single unified theory of sustainable development, all theories share a common theme in recognizing that future well-being is determined by what happens to wealth over time. Indicators of sustainability, therefore, tend to emphasize either stocks of wealth, or more specifically, how a portfolio of assets is managed over time. The contribution of the recent debate makes it clear that portfolio management must also take account of natural wealth, for example, changes in environmental liabilities arising as a result of pollution. It is on the relative importance of components of the portfolio of assets that opinions diverge into two broad camps. The first is 'weak sustainability' and states that total wealth should not decrease over time. Put another way, it is the 'overall' portfolio that is bequeathed to the future that matters. On this view, there is nothing special about natural wealth in so far as its liquidation is accompanied by adequate and offsetting compensating investment in other assets (Gibby, 1993).

In contrast, 'strong sustainability' suggests a greater emphasis on the conservation of natural assets within the broader goal of prudently managing a portfolio of assets over time. For advocates of strong sustainability, it is the physical protection of absolute levels of natural assets that is a prerequisite for sustainability. The rationale underlying this is that natural assets provide complex ecological functions crucial to the maintenance of life, and further, that these critical functions cannot be substituted for other assets (Norton and Toman, 1997). Natural assets that exhibit these characteristics are described by Pearce *et al.* (1989) as critical natural assets. It has been argued that allowing the stock size of these critical assets to fall below a threshold level could result in serious, if not catastrophic, consequences for well-being (Pearce *et al.* 1996). However, the problems involved in, firstly, identifying critical assets, and secondly, determining thresholds are far from trivial and are at the frontier of interdisciplinary research (Hamilton, 1997). Yet, some progress is being made in developing decision-making rules such as Safe Minimum Standards (SMS). An SMS is defined by Farmer and Randall (1998) as representing the suspension of standard arrangements for managing a given resource in order to prevent irreversible outcomes providing that this policy switch itself does not incur intolerable costs.

The predominant approach within the corporate environmental accounting literature is to take a definition of (usually, strong) sustainability from the broader literature and redefine it in this specific context. For example, Hawk en (1993) argues that a firm is behaving sustainably if it does not reduce the capacity of the environment to provide for future generations. Similarly, Bebbington and Gray (1997) assert that, at a minimum, a sustainable business is one that leaves the environment no worse off at the end of each accounting period than it was at the beginning. At first glance, these definitions are enticing, although somewhat stringent, serving both to focus attention on responsibility for environmental impacts and to suggest that firms should be more responsive to these concerns. In addition, Bebbington and Gray (1997) also note the trend by groups such as the Business Council for Sustainable Development towards avowedly pragmatic programs of action intended to be economically sound while decreasing environmental impacts.

## Major Issues in Environmental Valuation and Accounting

In environmental accounting, impairment of assets is a major issue. Assets are considered to be impaired when an organization is unable to fully recover the carrying value of the assets over their useful lives. This condition can result from inadequate depreciation charges, changes in production, competition or obsolescence. It can also result from contamination by environmental conditions. Normally, the accounting issues are:

- When to recognize impairment;
- How to recognize such impairment; and
- What disclosures are needed relative to the impairment.

Elements of accounting relative to environmentally impaired assets are:

1. Cost or market valuation procedures for inventory will recognize the written down value on the income statement.
2. Inventory is not written off if, in later years, the inventory recovers its value.
3. Temporary investments on temporarily impaired assets are written down but are written up to the original cost if they recover their values. The write-downs and write-ups are reported on the income statement.
4. Long-term investments are written down if the environmentally caused decline is believed to be temporary with a contra account adjustment in the equity section of the balance sheet. If the asset recovers its value, the entry is revised.
5. If the long-term investment has experienced a permanently environmentally caused decline in value, the loss is reported on the income statement as a loss. The determination of the permanency of the loss and the amount of the loss is a judgment call.

The criteria to be used for determining if an asset has been impaired are:

- **Economic** – If the carrying value of the asset is greater than the measured attribute of the asset, impairment has occurred and the asset should be written down.
- **Permanence** – If the above condition prevails and the impairment is thought to be permanent, then impairment has occurred and should be recognized in the accounts.

- **Probability** – If the above condition has probably occurred, the impairment should be recognized in the accounts. "Probable" is usually interpreted as more likely than not. It is not clear yet as to whether, if the asset value recovers, the write-down should be reversed and the asset returned to original cost. The write-down can be applied to a major group of assets, a subgroup, or to individual assets.

The handling of environmental remediation costs is another major issue to be addressed in the parlance of environmental valuation. Environmental remediation costs are the costs of cleanup of contaminated real property.

- The expenditure is incidental.
- There is no material addition to value.
- There is no extension of useful life.
- The purpose of the work is keeping the property in a normal efficient operating condition.

The Taxation Departments of most of the governments refer to the above identified items as various types of costs. Other related costs are: Assessment costs (the cost necessary to determine if a property is contaminated and to what extent). Normally, assessment costs are to be capitalized to the asset and are depreciated over its remaining useful life. In the same manner, remediation expenditures are also to be capitalized, especially where there is a plan in effect.

Even environmental audit and research and development expenses could be treated in the same way.

In US, the Clean Air Act of 1990 created an unusual asset, an emission right. The emission right is in effect a licence to pollute at a limited rate. The Act not only established the rights, it also set up the administrative machinery for the sale of these rights by one organization to another. The sale can be made by auction by the Environmental Protection Agency (EPA) or directly by the seller through the EPA.

The emission right is the authority to emit pollution at a specified time in the form of sulphur dioxide, a by-product of the burning of fossil fuels (coal and oil).

The unit of measurement is the amount of one ton of pollutant in a given year. Organizations that modernize and improve their power-generating systems, thus reducing the amount of sulphur dioxide emitted, can sell this excess pollution as available emission to another organization. This arrangement is important because the amount of sulphur dioxide cannot, by law, exceed the amounts emitted during a particular period. New plants (built after 1987) are not granted such rights. They have to buy them from existing plants.

The accounting question is how to treat these rights. Should they be treated as an inventory, as a marketable security, or as an intangible? It would appear that the first choice would be as an inventory item. The rights exhibit the characteristics that conform to items of tangible personal property which are held for sale in the ordinary course of business or are to be currently consumed in the production of goods or services. Although the rights are not tangible assets, they are as tangible as certificates of deposit and they do have value as related to future production. As a matter of fact, without the permission granted by these rights, a new plant could not produce and thus will not earn revenue.

The concepts of a marketable security or an intangible have some appeal because of the 'piece of paper' and the non-physical aspect of the rights. The latter concept probably holds more appeal because other types of intangibles do have some operating characteristics such as prepaid rent and prepaid insurance. There has been no definitive pronouncement on this subject.

Another problem that has not been definitive is the method of valuation of the rights. Should the method be cost or market? Cost of a right or allowance that is granted to an existing organization is zero if its omission is below the maximum allowance without any modification of its production operations. If the organization has installed improvements that reduce the output of sulphur dioxide and thus reduce the need for the rights by the performing organization, the cost of the improvements can be the basis for the value of the right. If the rights are purchased by another organization that needs them, the rights would be valued at the purchase price. If the rights are for future periods, discounting could be exercised.

Market value also could be available as it is assumed that there will always be a futures market for the rights. Also, the purchasing organization can value the rights based on its value to the production of its energy cycle. The selling organization also has this option. Hence, the potential selling price will compete with its economic value to production. The whole process is still unclear.

Valuation norms are applicable to environmental liabilities too. The determination of the liability for environmentally toxic conditions of property is an important aspect of both internal and external audits. Parties who originally placed the contamination on or in the property are responsible for the cost of cleaning up the contamination. The responsibility can also fall on successive owners of the property, even bankers, who, in the course of events, lent the money using the contaminated property as collateral. Thus, an owner who had no part in the action that contaminated acquired property may still be liable for cleanup costs.

Estimating cleanup liabilities is very difficult, and like all estimates, the amounts may vary. Some of the estimating problems are:

- The complexity of the sites.
- Potentially unreliable early estimates.
- The definition of acceptable cleanups.

Some of the activities relative to the cleanup operations are:

- Analyzing the environmental program.
- Hiring experts to assist in the study.
- Removing toxic waste and containers and disposing thereof.
- Sampling and testing soil.
- Neutralizing toxic wastes from the area concerned.
- Preventing future contamination [5].
- Putting the property back in usable condition.
- Monitoring the process.
- Estimating the costs from litigation.
- Estimating the costs imposed by regulatory agencies.

These costs can be best estimated by an environmental engineering or consulting organization. It is probably a good idea to estimate the costs at three levels—the best (lowest) cost estimate, the worst (highest) cost estimate, and the most probable cost estimate.

These valuation needs, therefore, call for an environmental audit program to be followed by the companies.

Despite the ongoing debate, there is some consensus that a prerequisite to understanding sustainable development is the construction of green accounts and sustainability indicators. At the macro level, green national accounting has been proposed, while at the micro or corporate level, considerable attention has been devoted to corporate environmental accounting and reporting. The latter covers a diverse range of activities with, currently, little standardization of existing practice (United Nations, 1997). If comparisons of environmental performance are to be made across the corporate sector, this will require an overarching framework within which to evaluate these activities. For example, within the domain of green national accounting, the United Nation's Satellite Environmental and Economic Accounts (UNSEEA) serve this function. This, in turn, is an adjunct to the conventional UN System of National Accounts (SNA). The SEEA framework embodies natural resource accounts, resource and pollutant flow accounts, environmental protection expenditure accounts and the estimation.

The control, the thrust area of this article, depends on environmental auditing by external and internal auditors which acts as a safety valve—the control that helps to ensure that there will be a minimum exposure to serious problems and then when the problems occur, the requirements of external agencies are met.

## Environmental Auditing

A competent environmental auditing program, conceivably a combination of internal and external auditing, is an excellent means for minimizing organizational environmental risk. This audit activity can be a part of the strategy of self-regulation. Audits can also be seen as a method of facilitating the disclosure requirements of governmental units. The audits determine the disclosures to have been or are being made and they validate the information that is being disclosed.

Like financial audit, environmental audit is being performed today by both internal and external auditors. Both are interested in many of the same areas of operation of the organization, but for different reasons. The internal auditor, as a part of the management and internal control fabric of the organization, is interested in:

- reviewing the compliance with regulations and statutes,
- determining the propriety of the accounting for environmental issues, and
- ensuring that proper disclosure is being made.

However, the internal auditor is also interested in the controls that are in place or that should be in place:

- to ensure that environmental problems are kept to a minimum,
- that environmental operations are efficient and effective,
- that waste management operations are proper, and
- that environmental decisions are based on factual information.

The external auditor is essentially interested in ensuring that the financial statements are proper, but must review many of the same elements of the organization's environmental operation.

The audit teams must be qualified for the operation. Thus, in addition to personnel with audit experience and ability, the teams should have qualified environmental engineers available on a full-time basis or as advisors when needed. The audit teams should also have legal expertise available as needed (Loren, 2000).

Though the teams function similarly, there should be independence and objectivity in their operation. The members of the internal audit team should be free from any organizational constraints and should have a direct reporting path to the top management.

Also, the audits should be performed in a systematic and well-organized manner. Audits should be well-planned and the audit programs should be a logical guide based on comprehensive preliminary survey processes. They should be well-documented and all elements of the audit reports should be supported in the working papers.

Internal auditing has two major customers: The auditee, the operating organization whose functions will be enhanced by the results of the audit; and the management that needs intelligence on the operating units for which it is accountable. In order to accomplish these dual responsibilities, the audit operations should cover these areas:

- **Compliance Audits:** Historically, this has been the prime function. As the audit horizon is broadening, this aspect becomes relatively less important.
- **Environment Management System Audits:** These audits determine how well the environment management systems are performing. Many authorities blame most environmental failures on deficient management systems.
- **Issue Audits:** These audits approach the audit from a functional standpoint. Issues of waste management, ozone depletion and topical deforestation are examples.
- **Preacquisition Studies:** These audits review organizations that are the target of mergers and takeovers to determine whether they are free from environmental vulnerabilities.
- **Supplier Audits:** These audits refer to the environmental performance of vendor and service suppliers such as waste disposal contractors.
- **Insurance Audits:** These are audits by insurance carriers to assess the environmental risk of their clients.

Attempts by the any external agency to force the conduct of internal audits of environmental operations have had a difficult time even in the USA (the agency being Environmental Protection Agency, commonly known as EPA), primarily because of the EPA policy of using these audits to determine infractions. The industry has resisted such attempts by stating that the internal audit is more an internal management tool and is a means of internally determining compliance.

External auditors must be familiar with the environmental aspects of reviewing assets and liabilities to determine:

- that the valuation is proper;
- that contamination has not reduced the carrying value of the assets, and

- that the expending and capitalization of remedial costs have been recorded properly.

External auditors will also determine that financial statements reflect the liabilities of the organization. Here, the auditors must, either by themselves or with environmental experts, determine that the organization is complying with governmental regulations in the handling of emission of pollutants, the disposition of contamination and waste and the detoxification of previously contaminated assets.

In addition, the external auditor must determine whether the client, as a result of the acquisition of new properties and assets, may be exposed to or actually have incurred liabilities as a result of the contamination of the acquired assets. The external auditor must also determine that proper disclosure in compliance with legal regulations has been made.

## Environmental Audit Techniques and Strategy

The strategy for conducting an environmental audit is much like the strategy for conducting ordinary audits, but the elements of the strategy are unique to the environmental area. It is to be presumed that both internal and external audit functions are well-organized and that the audit staff is knowledgeable about the normal audit operations. Briefly, the elements of the strategic approach include:

- **Objective:** The objective should be clearly stated—whether it is to determine compliance, examine operations to establish methodology propriety, determine costs, set up potential liability or whatever.
- **Personnel:** Based on the above objective, the audit team qualifications should be determined. This could include knowledgeable auditors, engineers, scientists and legal talent.
- **Audit Scope:** The scope of the audit as to the portions of the organization to be covered should be determined.
- **Documentation:** Information of government regulations and statutes, industry and government environment standards and state-of-the-art technology for operations and for environmental aspects pertaining to the organization should be collected.

- **Plans:** There should be plans established for the audit which should include: Interaction of the various disciplines, the degree of testing and scientific exploration that will be necessary, the methodology that will be used for cost estimating, the techniques of audit control that will be used, and the need for engaging specialists from outside the audit organization.
- **Audit Program:** There should be a plan for a preliminary survey and the resultant audit program.
- **Scheduling:** The segments of the audit should be scheduled with regard to time requirements and sequence.
- **Cost Estimates:** Costs of conducting the audit should be estimated and arrangements made for funding.
- **Criteria:** Criteria for materiality, sampling precision and other qualitative elements of the audit should be established.
- **Plans for Review:** Plans for review by the auditee, by legal experts, by engineering and scientific experts, and by the management should be set. Methods of resolution of differences should be set and agreed to.
- **Report:** The report format should be set. Also, plans should be made for establishing a rough outline of the report content as soon as the preliminary survey has been completed.
- **Distribution:** The distribution of the audit report should be established and approved by the audit specialists and by the management.
- **Media:** Plans should be made for relations with media representatives. There should be a single spokesperson appointed and adequate notice given to all other parties to keep a strict silence.

Each of the above areas should have contingency plans that will go into effect if the original plans are jeopardized and must be altered materially. The director of internal auditing is the normal contact and coordinator of the external audit operations and should coordinate with the external auditors as to the appropriate area of the strategic plan.

## Categories of Environmental Audits

There are several categories of environmental audits. They have both internal and external objectives. Internal objectives include informing management whether:

- Operations are in compliance with regulations;
- Liability is associated with property transfers;
- Contract waste management operations are being performed competently;
- Environmental management decisions are being made on the basis of facts;
- Environmental liability accruals are appropriate;
- Other activities that are being conducted are consistent and appropriate.

External environmental audits provide assurances to outsiders relative to the information relating to environmental activities that are included in or are a part of the usual financial statements (or that should be).

All environmental audits can be placed into seven classes. These audits are most likely to be performed as internal audits, or can be contracted for by outside specialist organizations, or can be performed by external auditors either in their own interests or for the internal auditors. The classifications are:

1. **Compliance Audits** determine whether or not activities and operations are within the legal constraints imposed by regulations. They are detailed site-specific assessments of current, past and planned operations. This classification includes three types of audit activities:
   - Preliminary assessments that are used to provide insight into potential problem areas, especially where future audit activity should take place.
   - Environmental audits that are more detailed audits focusing on operations. They include verification of compliance with permits and consent orders.
   - Environmental investigations or site assessments that are time and labor-intensive assessments, conducted when the preceding phases indicate that there is a risk of contamination or other non-compliance. They include interpretation of technical analyses.

2. **Environmental Management System Audits** focus on the systems in place, to ensure that they are operating properly to manage future environmental risks. The audits are conducted internally when the environmental auditing process matures.

3. **Transactional Audits,** also called acquisition and divestitive audits, property transfer site assessments, property transfer evaluations, and due diligence audits. They are a risk management tool for banks, land buyers, lending agencies and for any organization purchasing land for a facility site or accepting it as a gift or donation. The potential liability connected with an acquisition can easily exceed the market value of the asset.

4. **Treatment Storage and Disposal Facility (TSDF) Audits** involve the tracking of hazardous substances throughout their existence. All hazardous materials must be tracked from cradle to grave (creation to destination) and all 'owners' of these materials have liability for them as long as the owners exist. The liability accrues for ownership, leasing, managing or for loaning funds.

5. **Pollution Prevention Audits** are operational appraisals that serve to identify opportunities where waste can be minimized and pollution can be eliminated at the source rather than controlled at the end of the process.

6. **Environmental Liability Accrual Audits** are technical legal reviews involved with quantifying and reporting liability accruals for known environmental issues.

7. **Product Audits** are appraisals with the production process of a facility. The objective is to provide assurance that the product is in compliance with restrictions and with environmentally sensitive interests.

These audits are not exclusively performed, nor are all of them performed by internal or external audit staffs. They can be performed individually or in combinations. They are of use to operating units, to top management and in the preparation of financial statements.

Both internal and external auditors should review internal controls as related to environmental operations. Their review should focus on control procedures over the generation, transfer, elimination and cleanup of toxic substances.

These internal control audit procedures should extend themselves to the consideration and creation of liabilities for potential costs related to possible environmental agency sanctions, or for costs that will be incurred in cleanup and other remedial actions.

Both sets of auditors should hold discussions with the management and ultimately with legal counsel, internal and external, covering these controls and potential litigation, claims and assessments related to environmental activities and situations. Both audit groups should review legal fees with the intent of determining if there are any fees that are related to litigation or to other related environmental activity. If so, the auditors should fully investigate the situation and, based on this investigation, ensure that proper liabilities are established. In this connection, the auditors should determine the impact of such liabilities on the organization's resources as to claims in these resources. Correspondingly, there should be a consideration as to the entity's ability to continue as a going concern for a period of one year beyond the next balance sheet date.

## Environmental Risks and Exposures

The environmental risks and exposures to which internal and external auditors are subjected to may be enumerated as (Thompson and LeGrand, 1993) follows:

- Does the organization use toxic materials in its production operation?
- Does the organization, in its operation, emit hazardous materials into the environment?
- Has the organization operated, or does it operate motor pools or similar operations that could contaminate the ground?
- Does the organization's manufacturing operation create hazardous waste?
- Does or has the organization earlier operated on-site storage operations of hazardous supplies, materials or waste?
- Does the organization monitor the service contractor that disposes of hazardous waste or materials?
- Has the organization had environmental studies made of properties or organizations that it has acquired?

- Has the organization loaned money to other organizations that might be at environmental risk?
- Has there been a violation of any of the federal, state, or local statutes or regulations other than those identified above?
- Is there a possibility that the organization might be designated as a Potential Responsible Party (PRP)?
- Have actual and potential liabilities been identified and properly computed?
- Does the organization have an adequate environment management operation?

  (Current literature contains articles that describe in detail the requirements for such an operation.)
- Is there a potential for personal liability action?
- Is there documented evidence of the possibility of environmental problems that have not been reviewed?
- Is there a possibility that adjacent properties may be contaminated to the degree that there are effects on property that has been acquired or is being considered for acquisition?

The above list probably is not a complete list of the risks and exposures. The auditor should be alert to similar types of situations, and schedule for review those that are threatening. A method of establishing the priority of risks would be to assign an index for materiality and an index for probability. The combination of these indexes can establish a priority measure. However, it should be recognized that some risks would require mandatory prompt attention.

## Management's Ecological Responsiveness

Green accomplishments, competitive advantage, and financial impact of environmental audit are the three issues of an organization that wants to go for an environmental accounting program. In searching for perceived connections between a company's environmental accomplishments and a positive payoff in its competitive position, relative cost structure, product quality, reputation with customers, ability to compete in international markets and the development of unique or inimitable competitive advantages are to be analyzed.

As governments, businesses and international agencies raise the banner of global ecology, environmentalism changes its face. In part, ecology— understood as the philosophy of a social movement—is about to transform itself from a knowledge of opposition to a knowledge of domination (Sachs, 1993). Over the past decade, companies have begun to transform the environment from an operating problem into a business opportunity. As markets globalize, the challenge of 'sustainable development' has emerged as an important challenge for multinational corporations intent on expanding into emerging markets.

In the marketplace of ideas, the discourse of ecological modernization has been popular of late (Harvey, 1996; Hajer, 1996). In its most general form, the discourse of "ecological modernization depends upon and promotes the belief that economic activity systematically produces environmental harm and that society should, therefore, adopt a proactive stance with respect to environmental regulation and controls" (Harvey, 1996). However, it is this generality which has allowed a variety of organizations and institutions, with very different interests and objectives, to utilize the notion of ecological modernization to justify or rationalize particular interests as being in the global interest (Harvey, 1996). Much like the related notion of 'sustainability', ecological modernization has been used by groups as diverse as transnational capital, NGOs and grass-roots environmental organizations to argue for particular visions of the future.

Environment-related measures should demonstrate certain qualities for effective understanding and application. James (1994) lists eight criteria, which are as follows:

- **Comprehensible**—Measures should be simple and easy to understand.
- **Comprehensive**—A balanced scorecard is necessary, as no one measure can encompass all environmental concerns.
- **Comparative**—To place results in perspective and to provide benchmarks, comparability should be ensured.
- **Controllable**—Measures should be explicitly related to actions specified for improvement.
- **Credible**—Measures must be credible with key stakeholders, and must be seen as fair and feasible while focusing on key improvement areas.

- **Customer-focused**—Every measure must clearly define customers whose needs are taken into account in design and implementation.
- **Continuously Improvement-oriented**—Measures must highlight areas where improvement is necessary and feasible.
- **Cascading**—Broad measures must be derived from strategic business objectives and analyzed into sub-measures appropriate for different levels and activities.

The corporate policy makers in the third world as well as in the developed nations should take into consideration the following parameters while strategizing on environmental frontiers:

1. **Cost-effective Results**—Cost effective resources throughput and current or potential environmental burdens.
2. **Financial Returns**—Eco-efficiency metrics should transcend the central financial measure of return on investment [ROI] and reflect a life cycle costing perspective for integrating economic consequences of externalities into management decisions through 'whole systems accounting'.
3. **Environmental Efficiency**—Assessing environmental quality requires evaluation of many disparate factors. Some environmental factors that can be accounted for are direct costs and benefits of energy and material use, production efficiencies such as production cost reductions, environmental burdens such as toxic release inventory [TRI], and regulatory compliance such as compliance and non-compliance costs from eco-efficiency initiatives. These factors should be evaluated on a life cycle basis as far as possible.
4. **Productivity**—Natural resource productivity should be measured alongside productivity gains from labor and capital. Activities should be measured to show whether they squeeze more service per unit of resource consumed (i.e., reduction of throughput) instead of running more resources through the system.
5. **Customer Perception**—Impact of resource efficiency programs on customer loyalty should be considered to highlight marketplace benefits from sustainable business operations, since increasing market share surely remains an important managerial goalpost (Chatterjee, 2001).

To attain environmental sustainability, modernization of the intellectual application is needed, and as a result, companies that want to survive in the long run should be giving more thought to different environmental valuation techniques and accounting principles. At the same time, to achieve efficiency, principles of environmental audit may be implemented in these companies.

*(Arup Choudhuri is a Associate Dean, the Icfai Business School, Kolkata.)*

## References


1. Atkinson, Giles (2000): "Measuring Corporate Sustainability", *Journal of Environmental Planning and Management,* March, 2000, Vol. 43 Issue 2, pp. 235-253.
2. Atkinson, G, Dubourg, WR, Hamilton, K, Munasinghe, M, Pearce DW and Young, CEF (1997): *Measuring Sustainable Development: Macroeconomics and Environment* Edward Elgar: Cheltenham.
3. Bebbington, J and Gray, R (1997):" An Account of Sustainability: Failure, Success and a Reconceptualization", in: *Proceedings of the* 5th *Interdisciplinary Perspectives on Accounting Conference,* vol. 1, 7-9 July.
4. Bonifant, BC, MB Arnold, and FJ Long (1995): "Gaining Competitive Advantage through Environmental Investments," *Business Horizons,* July-August 1995, pp. 37-47.
5. Brown, Warren B and Necmi Karagozoglu (1998): "Current Practices in Environmental Management", *Business Horizons,* July/August 98, Vol. 41 Issue 4, p. 12-19.
6. Canadian Institute of Chartered Accountants (CICA) (1997), *Full Cost Accounting from an Environmental Perspective,* Canadian Institute of Chartered Accountants. Toronto.
7. Chatterjee, Kanika, "A Third Wave View of Sustainable Value Creation for Corporate Strategic Advantage", Strategic Cost Management, edited by AK Basu, S Banerjea and DR Dandapat, University of Calcutta, 2001.
8. De Koning-Martens, W and Van der Ende, N (1997) "Bridge Over 'Troubled Waters' or Chinese Walls': What Lessons can be drawn from the EMAA Workshop?" in: C Hibbitt and H Blokdjik (Eds) *Environmental Accounting and Sustainable Development:* The Final Report. Limperg Institute Amsterdam.
9. Everett, Jeff; Neu, Dean (2000): "Ecological Modernization and the Limits of Environmental Accounting?" *Accounting Forum,* March, Vol. 24 Issue 1.
10. Farmer, MC and Randall, A (1998): "The Rationality of a Safe Minimum Standard", *Land Economics,* 74(3), pp. 287-302.

11. Gibby, D J (1993): " Deductibility of Environmental Remediation Costs", *Journal of Accountancy,* December, pp. 44-9.
12. Hajer, Maarten (1996). "Ecological Modernization as Cultural Politics", in Lash, S, B Szerszynski, andB Wynne (eds.), *Risk, Environment and Modernity: Towards a New Ecology.* London: Sage: 246-68.
13. Hamilton, K (1997): Preservation and Strong Sustainability, *Environment and Development Economics,* 2(1), pp. 72-76.
14. Harvey, D (1996). *Justice, Nature and the Geography of Difference:* Blackwell: Cambridge, MA.
15. Hawken, P (1993): *The Ecology of Commerce,* HarperCollins. New York.
16. Kiernan, Mathew J and Jonathan Levinson (1997): "Environmental Drives Financial Performance: The Jury is in", *Environmental Quality Management,* Volume: 7, No.2, Winter.
17. Loren, Flo (2000): *Environmental Audit Handbook,* Stuart-Gill, Sydney.
18. Mookerjee, K (1998) "Coming Clean with Life Cycle Stewardship: The Way Beyond Corporate Governance" in *Studies in Finance and Accounting* edited by A K Basu, K Mookerjee and S Chakraborti. Calcutta: Department of Commerce, University of Calcutta.
19. Newell, G E, Kneuze, J G and Newell, S J (1992)., "Environmental Contamination", *The National Public Accountant,* June 1992, pp. 34-5.
20. Norton, B G and Toman, M A (1997): "Sustainability: Economic and Ecological Perspectives", *Land Economics,* 73(4), pp. 663-68.
21. Parker, L D (1997):" Accounting for Environmental Strategy: Cost Management, Control and Performance Evaluation", *Asia-Pacific Journal of Accounting,* 4(2), pp. 145-174.
22. Pearce, D W, Markandya, A and Barbier, E (1989): *Blueprint for a Green Economy,* Earthscan, London.
23. Pearce, DW and Newcombe, J (1998) *Corporate Sustainability: Concepts and Measures, mimeo* (University College London/University of East Anglia, Center for Social and Economic Research on the Global Environment (CSERGE)).
24. Sachs, W (1993). *Global Ecology: A New Arena of Political Conflict.* London: Zed Books.
25. Thompson, R P, Simpson, T E and LeGrand, CH (1993), "Environmental Auditing", *Internal Audit,* April 1993, pp. 20-1.
26. United Nations (1997): Environmental *Financial Accounting Guidelines, Report Prepared* for the UNCTAD Intergovernmental Working Group of Experts on International Standards of *Accounting and Reporting* (ISAR) New York, UN.

| Environmental Accounting |
|---|
| The environmental responsibilities that are faced by many companies include:<br>1. Meeting regulatory requirements or exceeding those expectations.<br>2. Cleaning up pollution that already exists and properly disposing of the hazardous material.<br>3. Disclosing to the investors, both potential and current amounts and nature of the preventative measures taken by management. (The requirement for disclosure is that the estimated liability must be greater than 10% of the company's net worth.)<br>4. Operating in a way that environmental damage does not occur.<br>5. Promoting a company-wide 'environmental attitude.' |
| *Source: http://acct.tamu.edu* |

# 14

# Beyond Environmental Accounting
## The Triple Bottom Line Approach

*Bibhuti B Pradhan and Sanjib Pattnaik*

***This article explores the advancement of the concept of corporate responsibility and a continuation of the discussion on Triple Bottom Line (TBL) as an approach in achieving corporate sustainability. It presents the path to sustainability that firms look for and work out strategies to guarantee financial success, while at the same time managing its environmental and social responsibilities. The notion of sustainability has grown to encompass economic, environmental and social dimensions. Today, the primary concern of many organizational managers is how to bring together these fundamental, yet seemingly disparate, pillars of sustainability together to form the integrated 'TBL' of economic, environmental and social performance. Ultimately, this article seeks to find a new set of management strategies and tools through TBL approach for managing and balancing corporate responsibility.***

*Source: Accounting World, January 2007.* 

Different companies view the concept of corporate responsibility differently. However, in most corporate discussions, corporate responsibility is categorized into three dimensions: Economic, environmental and social, where, corporate responsibility is translated into doing business profitably, ethically, and with sustainability.

Corporate responsibility is about addressing these economic, environmental and social responsibilities and managing them accordingly towards the attainment of a desired sustainability performance. These corporate responsibilities, therefore, could be translated into three sustainability pillars or the TBL, as shown in Figure 1.

**Figure 1: Corporate Responsibility Elements**

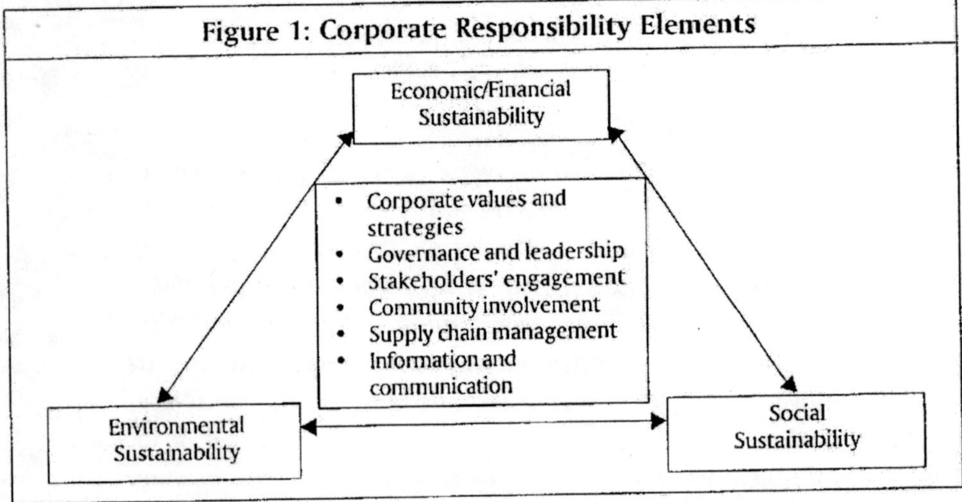


- Economic sustainability—economic profitability, competitiveness, job or market creation.
- Environmental sustainability —efficient use of natural resources and efficient environmental management and protection.
- Social sustainability—well-being of the society as a whole.

The path to sustainability requires that firms seek and work out strategies that guarantee financial success and at the same time managing the environmental and social responsibilities. Hence, the TBL approach looks at how corporations manage and balance all corporate responsibilities for a sustainable business, when most companies are facing such framework for integration. Although there has been proliferation of management systems, accounting, auditing and reporting standards, they focus on one or a combination of any two aspects of corporate responsibility. Hence, it is necessary to integrate these three responsibilities.

At the same time, as we are now operating in a 3-D world, the dimension of sustainability has surpassed all traditional concepts of environmental issues like population, global warming, biodiversity, etc. but we are closely moved with the sustainable capitalism transition of a global cultural revolution. This can be visualized from the following sustainable revolution's comparison of the new paradigm with the old paradigm.

**Figure 2: Seven Sustainable Revolution**

| S.No. | Criteria | Old Paradigm | New Paradigm |
|---|---|---|---|
| 1 | Markets | Compliance | Competition |
| 2 | Values | Hard | Soft |
| 3 | Transparency | Closed | Open |
| 4 | Life-cycle technology | Product | Function |
| 5 | Partnerships | Subversion | Symbiosis |
| 6 | Time | Wider | Longer |
| 7 | Corporate governance | Exclusive | Inclusive |

## Corporate Social Responsibility and Sustainable Development

Corporate Social Responsibility (CSR) is not a new concept. As early as 1953, Bowen (cited in Bucholtz, 1995) defined it as: "The obligations of businessmen to pursue those policies, to make those decisions, or to follow those lines of action which are desirable in terms of the objections and values of our society." CSR in the public sector refers to pursuing those policies, making those decisions and following those line of action that are desirable in terms of the objectives and values of society. (Armstrong, 2001). CSR as defined by the World Business Council for Sustainable Development (WBCSD, 1999) is "the continuing commitment by business to behave ethically and contribute to economic development while improving the quality of life of the workforce, and their families as well as of the local community and society at large".

A multifaceted concept of CSR includes intentions (moral or immoral), preventive and anticipatory actions, and social responsiveness (taking action, performance and accountability). CSR actions and responses refer to a wider range of activities that can include: Business conduct, environment, human rights, consumers, workplace, employees, community involvement and partnership, sponsorships or cause related marketing and philanthropy. The term is also used when referring to ethical investment or sustainability.

Ethical investment or Socially Responsible Investment (SRI), can be reactive, for example, not investing in such entities concerned with tobacco, alcohol, gambling, animal testing, nuclear weapons or pollution; or proactive, investing in companies demonstrating concern for the environment and supporting those that promote dissimilar values. Ethical investment is a well-established, major investment sector in the US, Europe and Australia. In the US, 13% of the $16.3 tn in investment assets is invested in socially responsible funds and they have consistently outperformed the market average (KPMG, 2001).

Sustainable development refers to more than concern for the physical environment. It is about managing environmental, social and economic outcomes for the benefit of all stakeholders. The priority areas that should be addressed for reporting sustainable development are human rights, employee rights, environmental protection, supplier relationships, community involvement and stakeholder rights. The mapping sentence suggests that accountability for duties and responsibilities to multiple stakeholders is demonstrated through Triple Bottom Line Reporting (TBLR), i.e., the reporting of financial, ethical/cultural/ social and environmental data.

The mapping sentence in Figure 3 suggests that accountability for duties and responsibilities to multiple stakeholders is demonstrated through TBLR, i.e., the reporting of financial, ethical, cultural, social and environmental data. The notion of sustainability has grown to encompass economic, environmental and social dimensions. Today, the claim of social and ecological market is expressed through the goal of sustainable development and there are many indicators that this will be an increasingly more important goal for the society (Save, 1996). At the corporate level, sustainability issues have become one of the main driving forces in running a business successfully.

**Figure 3: Mapping Sentence Describing the Relationship between Corporate Governance, CSR and TBLR**

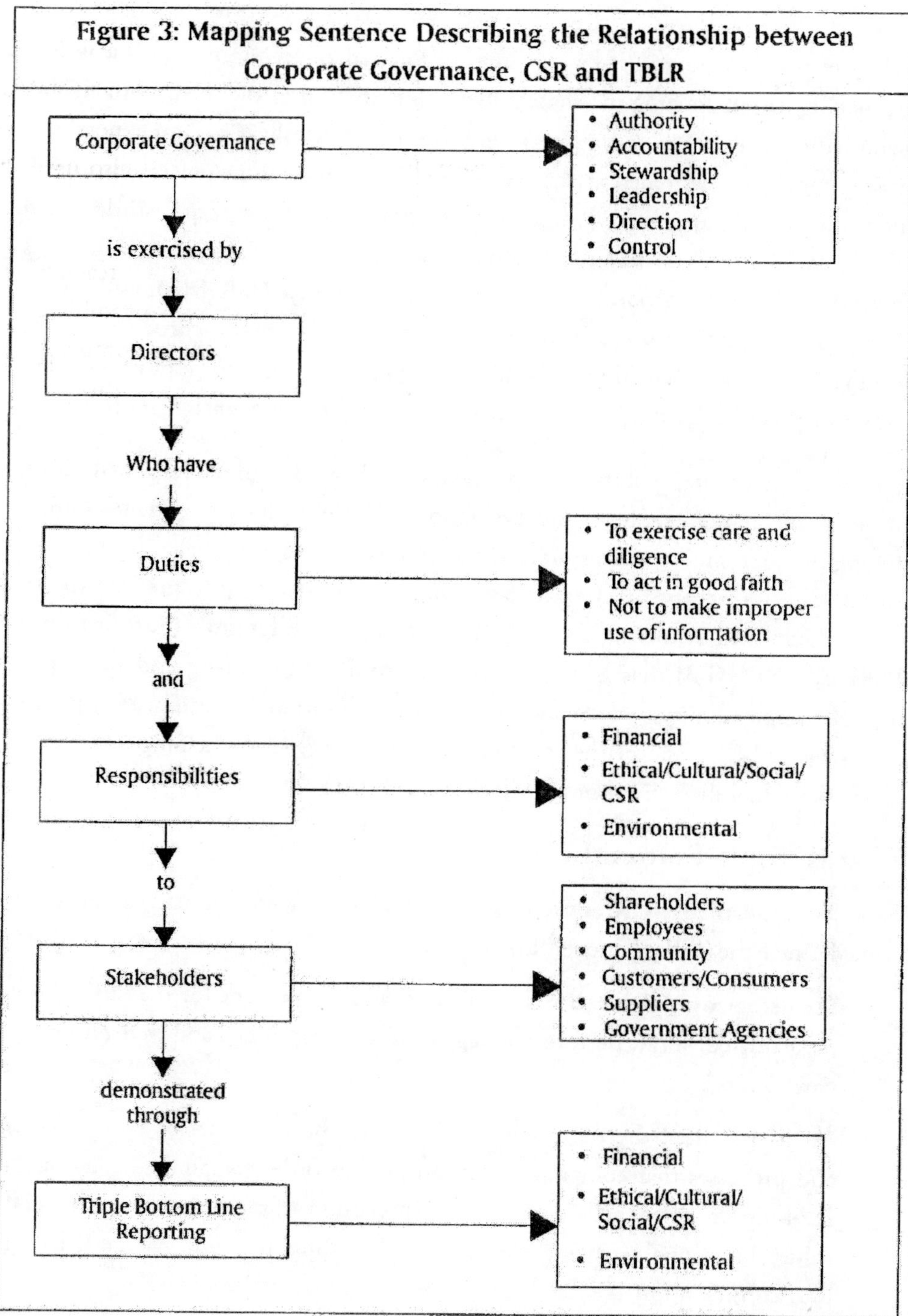

In the past, corporations have primarily focused on economic responsibility, i.e., making or maximizing the shareholder value, instead of the theory of a profit-maximizing firm. When the time of environmental revolution came about two decades ago and heightened at the Rio Conference, environmental responsibility (including health and safety) has tendered with economic responsibility and become the focus of corporate sustainability. Recently, social responsibility has emerged as the new corporate challenge to sustainability. CSR has now become the top list of new concerns in many businesses. Today many corporations are working towards a new goal of embedding social issues in the corporate strategies, ethos and practices as the latest strand of the fabric of corporate sustainability.

As a result of the evolutionary processes of these corporate responsibilities, various performance measures and evaluation techniques have proliferated, thus, gauging the companies' progress in terms of doing business in a sustainable manner. One of the latest approaches to sustainability performance measurement, management, and reporting is the TBL proposed by Elkington (Sustainability, 1998). Recently, TBL has become a popular tool in measuring and managing corporate responsibility performance in an integrated or holistic fashion. Corporate experts rate TBL as one of the most effective frameworks for helping companies achieve sustainability (Environics Globe Scan 2001).

## What is Triple Bottom Line?

John Elkington of the UK consultancy Sustainability, who originally coined the term, defined the TBL approach as:

- "At its narrowest, the term TBL is used as a framework for measuring and reporting corporate performance against economic, social and environmental parameters".
- "At its broadest, the term is used to capture the whole set of values, issues and processes that companies must address in order to minimize any harm resulting from their activities and to create economic, social and environmental value. This involves being clear about the company's purpose and taking into consideration the needs of all the company's stakeholders".

As promoted by the Center for Innovation in Corporate Responsibility (CICR), the concept of the integrated TBL of corporate sustainability represents, to date the most comprehensive approach to understanding corporate responsibility and bringing together the environmental, social and economic bottom lines of a business. It attempts to bring these three inter-related spheres of activity together in order to present a balanced view of overall corporate performance. Long-term corporate operational sustainability is essentially a function of how well a company can identify, understand, evaluate and manage its position within these three spheres—ignoring or undervaluing one or more of these spheres will eventually impact a company's ability to compete in an increasingly turbulent operating environment. The TBL can be communicated to all stakeholders describing the company's approach to managing one or more of the economic, environmental and/or social dimensions of its activities and through providing information on these dimensions, which can be popularly known as TBLR system.

Today, the primary concern of many organizational managers is how to bring together these fundamental, yet seemingly disparate pillars of sustainability together to form the integrated TBL of economic, social and environmental performance. Is measuring and assessing corporate performance in one or two, but not three spheres not sufficient? For open-minded managers, the answer is obvious: Each of these spheres is interconnected with the others. Operating environments become increasingly competitive and complex; it becomes essential to ensure that all spheres are working together in a complementary and reinforcing fashion. Progressive companies will realize that such 'good management' will help entrench and maximize the values inherent in each of the individual economic, environmental and social bottom lines (CICR).

Hence, there are some leading corporations that publish sustainability reports highlighting their economic, environmental and social performances. Some progressive and well-known international companies such as Novo Nordisk, WMC, Dow Chemical, DuPont, British Airway, BP Australia, Telstra, Shell have attempted with varying degrees of success to bring the three responsibilities together under one performance management, evaluation and reporting framework. International organizations such as the World Resource Institute (WRI), Coalition for Environmentally Responsible Economies (CERES), United Nations Environment

Program (UNEP), International Chamber of Commerce (ICC), Global Reporting Initiative (GRI), and World Business Council for Sustainable Development (WBCSD) are also actively contributing to the development of standards and processes within the TBL framework. These initiatives have established an important baseline of information regarding what works and what does not.

## TBL Approach

The TBL approach provides more information than the financial information on economic, environmental and social information as discussed earlier for sustainable business activities of the organization in response to mandatory requirements; consistency with emerging public communities by business through voluntary codes of behavior or charts; and the increasing demands from stakeholders for greater transparency about operating policies and results. These are in the form of inclusion of environmental and social information within annual reports to shareholders, a separate environmental report or community report, separate environment and social report, combined social and environmental report or full TBL report.

As TBL reporting usually contains both qualitative and quantitative information, it has to have the qualities and characteristics of information on reliability, usefulness, consistency of presentation, full disclosure, reproducibility, and auditability featuring as annual reports to bring objective, balanced and credible information of the reporting companies.

The following elements can be identified as core specific benefits to a TBL approach.

- **Broadening the basis of performance evaluation:** Amongst New Zealand TBL reporters to date, it is typically meant the expansion of annual reports to cover qualitative and some quantitative analysis of:
  - Environmental performance in production and operation.
  - Employment issues (training, turnover, staff, attitudes).
  - Equal opportunities and diversity practice.
  - Occupational safety and health.
  - Aspects of other stakeholder relations (e.g., customer satisfaction).

- **Philanthropic/community engagement initiatives:** TBLR reports to date have tended to specify social, environmental or economic objectives and to report performance against related commitments or targets. Indicators of progress are expressed in a mixture of quantitative metrics or even qualitative terms. Various companies in New Zealand have partially adopted the principles and content guidelines of the Global Reporting Initiative (GRI), as a basis for their reporting. The GRI seeks to promote consistency and comparability of 'Sustainability Reporting' by defining a minimum set of non-financial indicators and policy areas to be addressed in a TBLR type report. In most published TBL reports, there is also a tendency to emphasize reporting on 'internal' operations of an organization (office energy, staff attitudes, etc.), over the reporting of the 'external' impacts or value added from products and services on the environment and society. Understandably, this emphasis reflects the complexity of the issues surrounding 'external' activities.
- **A management process:** TBLR is about more than reporting, it is a management process covering strategic planning and objective/target setting; accounting for performance; auditing and reporting on the performance and embedding the results within organizational processes. This is generally set in the context of a process of continual improvement.
- **Stakeholder engagement:** TBLR includes the targeted identification and engagement with stakeholders of an organization in the process of performance evaluation. Stakeholder theory is based on the premise that the performance of an organization is reliant on more than meeting the needs of its legal 'stakeholders', such as a corporation's shareholders. The UK Institute of Social and Ethical Accounting AA 1000 Standard is generally acknowledged as the most developed tool to be used by corporations in their stakeholder engagement. This ensures the companies on attracting the different segments of investors to the market by reducing overall risk factors.
- **Increasing accountability and transparency:** The increased accountability and transparency of TBLR in the private sector can be seen as a response to the demands by NGOs and community groups for multinational corporations to justify their 'license to operate'. However, post-Enron/WorldCom the call for greater accountability and transparency has become a major public concern; thus, improving the corporate reputation and brand benefits.

- **Attracts and retain high caliber employee:** The publication of TBL related information not only satisfies the expectations of existing and prospective employee, but also plays a major role in enhancing employee royalty, reduces staff turnover and increases the company's ability to attract high-quality employees.
- **Innovation:** The development of innovative products and services with efficient utilization of resources and materials can be visualized through TBL reporting in alignment with research and development activities on the basis of stakeholders' priorities and concern.
- **Comparison of the bottom lines:** TBLR seeks to manage and report on performance across the social, environmental and economic dimensions across companies for the benefit of stakeholders. There is an ongoing debate/ discussion about:
  - The extent to which the social and environmental performance parameters can be quantified and compared with quantitative economic indicators.
  - Whether a common metric is the only basis on which to make meaningful decisions about trade-offs between the various bottom lines.

Debate abounds in the corporate world about the trade-offs between duties to ethical governance principles and shareholder returns. Some see the only responsibility of a board as that of making a profit for their shareholders and consequently abhor the costs and potential disclosure offered by TBLR.

## GRI Indicator Index

The GRI has committed significant efforts to the development of sustainability/ TBL indicators, which has gained broad acceptance on economic, environmental and social dimensions, which has been presented in Figure 4.

Others see business operating for the welfare of society as a whole and support the view that, as business is the third great institution of society after government and law, its social responsibilities encompass every aspect of business operations from the products companies make and the services they deliver to the relationships they have with employees, customers, suppliers and the communities in which they operate.

**Figure 4: GRI Indicator Index**

| Indicator Type | Category | Aspect |
|---|---|---|
| Economic Performance Indicators | Direct Economic Impacts | Customers, Suppliers, Employees, Providers of capital, Public sector |
| Environmental Performance Indicators | Environmental Impacts | Materials, Energy, Water, Biodiversity, Emissions, Effluents and waste Suppliers, Products and services, Compliance, Transport, Overall |
| Social Performance Indicators | Labor Practices and Decent Work | Employment Labor, Management/relations, Health and safety, Training and education, Diversity and opportunity |
| | Human Rights | Strategy and management, Non-discrimination, Labor management relations, Freedom of association and collective bargaining Child labor, Forced and compulsory labor, Disciplinary practices, Security practices, Indigenous rights |
| | Society | Community, Bribery and corruption, Political contributions, Competition and pricing |
| | Product Responsibility | Customer health and safety, Products and services |

External stakeholder groups, who have stimulated concern for public issues such as human rights, sustainable development, health and safety, consumer concerns, the rights of animals and the environment, are also demanding that business meet their corporate social responsibilities. Investors and consumers are taking a much greater interest in the reputation and credibility of corporations (Tomasic, 1993). The Good Reputation Index (Age, 2001) was the first index to rank Australian companies by reputation. The top 100 companies, selected for financial performance from BRW magazine's annual list of the top 1000 listed companies in Australia and New Zealand, were rated by representatives of 19 organizations on indicators of market position; financial performance; management, ethics and governance; social impact; environmental performance, and employees.

The company with the best reputation is ranked number 1. The ratings of performance are compiled from opinions of relevant stakeholders for each category, based on their perceptions of actual management practice. It appears that increasingly major investors such as the superannuating funds are taking these criteria into account when making investment decisions.

## Conclusion

Companies are becoming increasingly aware of the benefits of a good corporate reputation. The introduction of socially responsible reporting is a sensible thing to do, given the growing impact on investment decisions. There is a growing body of knowledge on measurement and standards for social and environmental reporting and agreement in the international community on the need to include this kind of reporting as a component of best practice corporate governance. Francis and Armstrong (2000) and others also demonstrate that there is a link between a corporation's ethics and its performance, and that is possible to be socially responsible as well as financially profitable.

In India, Tata Steel has gone ahead in environmental aspects, where RP Sharma, Head-Environment Division has rightly remarked, "At Tata Steel, we have always followed the proactive approach to protect the environment in and around the operating facilities of the company. In order to maintain our leadership we are now integrating Corporate Sustainability Management System (CSMS) with Tata Business Excellence Model which will address the stakeholder concerns in a systematic manner and we would be in a position to improve our TBL i.e., economic, environment and social performance.

*(Bibhuti B Pradhan, Professor and Dean, Research and Development Institute of Business and Computer Studies, Bhubaneswar. The author can be reached at bbpradhan@ibcsindia.com*

*Sanjib Pattnaik, Assistant Professor, Institute of Business and Computer Studies, Bhubaneswar. The author can be reached at sanjib_3435@yahoo.com).*

# 15

# Red Carpet for One-Stop-Shop Environmental Courts in India

*Lakshmi Lella*

***The multifaceted legal issues relating to science and technology are increasing day by day, leading to complex environmental litigations. The courts thus stand in a delicate position, as they have to choose between environment and economic development, while both are equally important for the development of the nation. For this very reason, the courts seek expert opinion which delays the adjudication of environmental matters. This problem could have been controlled, had there been a statutory panel of members comprising technical or scientific experts along with the judicial members to look exclusively, into environmental matters, as in other countries. Hence, this paper analyzes the pros and cons of constituting separate environmental courts in India, and advocates for the same. In this regard, this paper proposes to look into the need, circumstances, judicial opinions, and constitutional mandates for constitution of such special courts; despite the fact that there exist environmental benches and tribunals. It further analyzes the 186th Report of Justice M Jagannadha Rao,***

*Source: The Icfai Journal of Environmental Law, October, 2006.* 

***Law Commission of India, on the 'proposal to constitute environmental courts'[1] which propose for the constitution of separate environmental courts.***

## Introduction

The tight rope on which the Indian Judiciary has to walk while adjudicating the environmental issues has been highlighted by the Apex court, many a time. The Apex Court was repeatedly criticized for its non-technical application of mind; and on its part, the Court has repeatedly expressed its hardships, and restraints in its judgments, namely, *M C Mehta vs. Union of India,*[2] *Indian Council for Environmental-Legal Action vs. Union of India*[3], *AP Pollution Control Board vs. M V Nayudu,*[4] *and AP Pollution Control Board vs. M V Nayudu II*[5]. It expressed the need for constituting separate technical courts exclusively to deal with environmental issues. In addition to the pressure from judiciary, submission of reports by Malcolm Grant[6] and the Report of the Royal Commission[7] in UK, culminated in the appointment of the M Jagannadha Rao Commission.

The Jagannadha Rao Commission was entrusted with the study of the details regarding the constitution of environmental courts in India. These environmental courts are expected to be comprehensive, having both judicial and technical inputs as formulated by Lord Woolf in England[8], and something parallel to the existing environmental court legislations in Australia, New Zealand and other countries. The Commission submitted its report in two parts. The first part consists of a review of existing laws on environment courts in each State, and includes the suggestions and recommendations for the proposed courts. Hence, this paper intends to look into the various aspects of the constitution of these exclusive 'one stop-shop-environmental courts', and presents the case for early action in this regard.

## Judicial Opinion on Scientific and Technological Advances, and Environmental Issues

While the advances in science and technology in the last few decades have been extraordinary, so has been their impact on environment. The courts have to face multifaceted issues of science and technology, either directly or indirectly, raised before them, relating to the environment, especially, in matters concerning water and air pollution, cleansing of rivers, streams and lakes, or disposal or recycling of waste and sewage, toxic waste, hospital waste, nuclear waste, radio active material, removal of the effect of detergents, waste-oils, genetically modified organisms, adverse effects of pesticides, asbestos, etc. Air and water pollution caused by various industries like steel, textiles, leather, and traffic is severe too. The problems of climatic changes, depletion of ozone layer, and noise pollution—both at the workplace and in residential areas, are growing multifold. The problems of conservation and protection of wildlife, improper Environment Impact Assessment (EIA) system, etc. are some such problems faced by the courts.

The technical and scientific problems arise at various stages of setting up of an industry, and sometimes, at the initial stages itself. For instance, when an industry is established, and it installs the effluent treatment or cleansing mechanisms, or the pollution-prevention, or pollution reduction systems, the Pollution Control Boards object that they are not sufficient. Or when the industry and the Pollution Control Board agree that the safeguards are sufficient, the local public may object to the pollution caused by the industries as being unbearable. Issues like whether the running industry be ordered to close down or allowed to continue with better safeguards, can also arise. Hence, the courts are supposed to be ready with both short-term, as well as long-term remedies. An order of closure of industry may lead to unemployment for hundreds of employees, or result in loss of excise duty or sales tax to the government. In case of ordering shifting the industry, land at a subsidized rate, and shifting costs have to be provided by the Government to the industry. On the other hand, if the polluting industry is not shifted, the health and well-being of citizens residing in and around that area would be affected.

The courts, however, cannot order for the closure of an industry based only on the evidence produced by the industry itself. This is so, because at times, the industrialists themselves abuse the legal procedures for their own gain and want the industry closed down under the court orders, in order to escape from the statutory procedures for closure as provided in the Industrial Disputes Act, 1947; or to evade from paying compensatory wages to the workmen. At times, public interest cases are filed to blackmail the industrialist. Tracking all these issues is very difficult for regular courts due to their huge workload and the absence of technical experts. Further, in the name of prevention of pollution and environmental damage, the need for the development of industries, or irrigation or power projects should not be overlooked by the courts while pronouncing the judgments.

Considering all these factors, the Supreme Court has given balanced views in several cases and has directed shifting of the industries outside the cities. It had framed schemes for the payment of compensation for workmen, or for their re-employment or provided land for the industry elsewhere, but only with the help of expert opinion.

The US Supreme Court, in its landmark judgment in *Daubert vs. Merrel Dow Pharmaceuticals Inc.*[9] referred to the different goals of science and law in ascertaining the truth. The court observed that in the case of science, the quest for truth ends in the laboratory, and scientific conclusions can be subjected to continuous revision; whereas in case of law, the disputes must be resolved finally and quickly once and for all, leaving no chance for revision of any kind. Another distinction between science and law is that the scientists can improve, alter or abandon variables or models whenever more information is available, whereas it is not the same in case of courts as they have to choose from existing scientific knowledge. Consequently, environmental issues cannot be adjudicated by the courts merely by applying scientific variables, especially when scientific knowledge is stratified in policy-making, and the same has to be used as a base for decision-making by adjudicating agencies and courts.

The judge adjudicating an environmental issue has to keep guessing whether to proceed on the basis of doubts of an affected party, or to rely on the promises made by the polluter—and this is where the technical expertise is required. For instance, in the case of *Vincent vs. Union of India*[10] the petitioners sought a direction banning import, manufacture, sale and distribution of certain drugs as recommended by the Drugs Consultative Committee. The Supreme Court did not refer the matter to an independent scientific body and accepted the Committee's Report, stating that once the experts had approved or disapproved the drugs, the Court would not like to go into the correctness of their decision. Whereas, in *Dr. Shivrao vs. Union of India*[11], the Supreme Court referred the issue to an independent committee of experts. Similarly, in *AP Pollution Control Board vs. M V Nayudu*[12], the Apex Court proceeded to have the claims of the party tested by experts. The case of *AP Pollution Control Board vs. M V Nayudu*[13] is another example of the benefits of extensive scientific investigation. Had this scientific investigation not been carried out, the lives of millions of citizens could have been endangered. The precautionary principle was applied here.

Thus, it is well established that scientific conclusions are based on the 'data' and 'procedures' applied by the scientific institutions concerned. The conclusions are correct to the extent of the correctness of the data available or to the extent of the efficacy of the procedure and technology adopted for the analysis. With more data and application of better scientific procedures, or better technology, more accurate conclusions could always be arrived at which make any study tentative.

The High Courts and the Supreme Court of India have been taking up these and other complex environmental issues without an independent statutory panel of environmental scientists to help and advise them on a regular basis. In the absence of an expert panel, there is every likelihood of the Courts applying the Wednesbury Principle, which means not going into the merits of the case and deciding the same without conducting any spot inspections or adducing evidence to observe the facts as they exist on the ground.

## Environmental Issues and Existing Legislative Provisions

The Apex Court and its subordinate civil and criminal courts initially were empowered to exercise powers with regard to public[14] and private nuisances. Criminal Courts could exercise jurisdiction over offenses under various sections of the Indian Penal Code relating to environment[15]. The Criminal Procedure Code of 1973 in Chapter X (B) &( C) provides for the control and removal of nuisance. Presently, there are various legislations which provide for protection of natural resources like the Water (Prevention and Control of Pollution) Act, 1974; the Air (Prevention and Control of Pollution) Act, 1981; the Public Liability Insurance Act, 1991; the Forest (Conservation) Act, 1980; the Wild Life (Protection) Act, 1972; and the Environment (Protection) Act, 1986.

Any person whether directly affected or not, has the *locus standi* to raise an environmental issue before the High Court or Supreme Court. Nevertheless, the environmental matters are not taken up regularly on a day-to-day basis by the High Courts or Supreme Court, and hence result in delay of delivery of justice.

In order to hasten the implementation process, certain rules were framed over a period of time, empowering the quasi-judicial authorities with special powers to look into environmental issues. However, these bureaucrats, not being technical experts in the respective areas, are unable to do full justice. Hence, many appeals are pending before the Apex Court. The appeals lie to the Supreme Court from the following quasi-judicial authorities appointed under various rules:

- Authorities constituted under Section 25 of the Environment (Protection) Act and rules made under Section 3(3) of the Environment (Protection) Act, 1986.
- Member-Secretary, State Pollution Control Board or any officer designated by the Board to the Secretary, Department of Environment of the State Government under The Hazardous Wastes (Management and Handling) Rules, 1989.
- Authorities or persons constituted under Manufacture, Storage and Import of Hazardous Chemical Rules, 1989.
- Authorities constituted under Municipal Solid Wastes (Management and Handling) Rules, 2000.
- The appellate authorities like Secretary, Ministry of Environment and Forests, or in certain cases, the Deputy Secretary in the same Ministry under the Ozone Depleting Substances (Regulation and Control) Rules, 2000.
- The authority constituted under the Noise Pollution (Regulation and Control) Rules, 2000, and includes a District Magistrate, Police Commissioner or any officer not below the rank of a Deputy Superintendent of Police.
- The State Pollution Control Board in States, or such Committees in Union Territories constituted under the Bio-Medical Waste (Management and Handling) Rules, 1998.

## The National Environment Tribunal Act, 1995

The National Environment Tribunal Act, 1995 has a twin objectives: (i) To impose strict liability for damages arising out of any accident occurring while handling any hazardous substance[16]; and (ii) To establish a tribunal for effective and expeditious disposal of cases arising from accidents, and provide compensation for the losses incidental to such accidents.[17]

The tribunal consists of a chairperson, vice-chairpersons and judicial members and technical members as the Central Government deems fit under Section 9(1) of the Act. The tribunal is empowered to sit in benches each with a judicial and technical member. An appeal lies to the Supreme Court on a question of law from the award of the tribunal.

On the other hand, this Act also suffers from a few drawbacks. The tribunal is empowered only to award compensation unlike the civil courts which have wider powers such as permanent and mandatory injunctions, or orders for possession, etc. The members of the Tribunal need not be experts in the field of environment[18].

## The National Environmental Appellate Authority Act, 1997

The National Environmental Appellate Authority Act, 1997 was intended to constitute the National Environmental Appellate Authority to hear appeals with respect to restrictions imposed on any industry, whereby any operation or process shall or shall not be carried out, subject to safeguards under the Environmental (Protection) Act, 1986. The Appellate Authority consists of a chairperson, a vice-chairperson and such other members not exceeding three, as the Central Government may deem fit.

## Environmental Courts in Australia, New Zealand and the UK

Countries like Australia and New Zealand have taken the lead in establishing Environmental Courts, which are manned by Judges or Commissioners, having expert knowledge in environmental matters.

### Australia

In Australia, in the State of New South Wales, the Land and Environmental Court was established by legislation in 1980, under the Land and Environment Court Act, 1979. It is a superior court of record and comprises of judges and nine other technical and conciliation assessors. It has appellate jurisdiction, and is also endowed with judicial review and enforcement functions in relation to environmental and planning law. Proceedings can be commenced by anyone. Under this Act, a person may be appointed if he has experience and special knowledge in administration of local government or town planning; or in town or country planning or environmental planning; or in environmental science or matters relating to the protection of the environment and environmental assessment; or in the law and practice of land valuation; in architecture, engineering, surveying or building construction; in the management of natural resources or the administration and management of Crown lands, lands acquired under the Closer Settlement Acts[19]. The proceedings are to be conducted by the Commissioners with as little formality and technicality as possible, and expeditiously. Hence, the Court is not bound by the rules of evidence. However, it may obtain assistance of any person having professional or other qualifications[20].

## New Zealand

The New Zealand Environment Court was established under the Resource Management (Amendment) Act, 1996, by amending the Resource Management (Amendment) Act, 1991. It is an independent specialist court consisting of environment judges and commissioners, who are appointed for a period of five years. While appointing the judges and commissioners, the Governor General considers the need to ensure a mix of knowledge and experience—in commercial and economic affairs; local government; community affairs; planning and resource management; heritage protection; environmental science; architecture; engineering; minerals; and alternative disputes resolution processes. The duties of the Court include avoiding, remedying or mitigating adverse effects on the environment, in addition to the general duty to promote sustainable management. Further, the Courts are to encourage mediation and arbitration. The Court is not bound by rules of evidence and is free to establish its own rules of procedure. The proceedings are often less formal, and the parties can either appoint a lawyer or argue on their own. An appeal may lie to the High Court on the questions of law only.

## United Kingdom

In the UK, the Department of Environment was set up in November 1970. The statutes governing environment are the Wildlife and Countryside Act, 1981, the Water Act, 1989, the 1990 Planning Acts and the Environment Protection Act, 1990 (EPA). However, the credit goes to Lord Woolf, who advocated for the constitution of separate courts to deal with environmental issues. He expressed his support for the Environmental Courts and advocated that the tribunal must consist of architects, surveyors and a multi-disciplined adjudicating panel with broad discretion. He further suggested that these courts must be multifaceted and multi-skilled bodies, with a combination of services provided by the existing courts, tribunals and inspectorate in the field of environment. The courts must be a one-stop-shop with faster, cheaper and effective methods for the resolution of disputes, so as to avoid load on the already overburdened judiciary or other institutions that are compelled to resolve the issues within their minimal technical and procedural limitations. The judges in these courts must examine environmental problems with a broader vision, and apply all their technical knowledge and experience related to the environment in dispensing environmental issues. Hence, the present Environmental Courts in the UK have all the technical expertise to deal with environmental issues of all kinds.

## Need for Environmental Courts

The immediate factors that persuaded the Indian Law Commission to study the practicability of establishing separate environment courts are:

- Repeated emphasis by the Apex Court to constitute a separate tribunal with a panel of experts to deal with environmental issues, and to speed up the delivery of justice.
- Existence of the contemporary practice of taking expert advice from non-judicial members like environmental scientists or technically qualified persons as part of the judicial process.
- Drastic disparities in the constitution of appellate authorities under various legislations[21].
- Environmental matters being dispensed by bureaucrats on the bench with no judicial or environmental backup.
- Uneven, haphazard structure and functioning of quasi-judicial bodies and authorities[22].

- Absence of laboratories and other mechanisms to enable assessment of the scientific data by the regular courts.
- Delay in dispensing environmental issues by the Courts due to the lack of technical knowledge, and the tedious procedure of the courts.
- The ineffective work output by the National Environmental Appellate Authority constituted under the National Environmental Appellate Authority Act, 1997 due to its limited objectives and purpose.
- Lack of independent expert advice to the Bench, to handle the uncertainties of scientific conclusions.
- Pressure from within the country and outside, to strike a balance between sustainable development and economic development, between the closure of the polluting industries and rehabilitation of retrenched employees.
- The need to evolve a refined Neo-environmental Jurisprudence in accord with scientific, technological developments, international treaties, conventions and judicial decisions.
- To fulfill the Constitutional objectives of Articles 21, 47, 48A and 51A(g) by means of a fair, fast and satisfactory judicial procedure.

An alternative proposal of the government to constitute a single Appellate Court at Delhi over all the other statutory authorities, was not acceptable to many, as it would practically be impossible for a person or affected group residing in some distant place to go all the way to Delhi to raise his/her grievance. Another likely problem which could arise is the piling up of environmental cases before the appellate authority, as it is the only single appellate authority for the whole country.

## Recommendations of the Law Commission

Based on the views expressed by the judicial fraternity, the Constitutional validity, and the contemporary reports on constituting the Environmental Courts, the Law Commission recommended constitution of separate Environmental Courts in India too. It specifically recommended the following:

- **Composition:** Environmental courts are to be manned by persons having judicial experience, assisted by persons having scientific qualification and experience in the field of environment.

- **Object:** The Union Government must establish environmental courts in each State, in order to meet the objective of accessible, quick and speedy justice to the needy at their doorsteps.
- **Constitution of Members:** The environmental court to be headed by a chairperson, and assisted by two other members. The chairman and other members should either be a retired judge of the Supreme Court or High Court, or have at least 20 years experience of practice as an advocate in any High Court. Each environmental court is to be assisted by a panel of at least three scientific or technical experts known as Commissioners.
- **Jurisdiction:** The environmental court to have original jurisdiction in original civil cases where a substantial question of law relating to the environment, including the enforcement of any legal or constitutional right relating to the environment, is involved. The original jurisdiction of civil courts is not to be ousted, and the Environment court shall have appellate jurisdiction in respect of appeals, coming from the following:

  i. The Environment (Protection) Act, 1986 and rules made thereunder;

  ii. The Water (Prevention and Control of Pollution) Act, 1974 and the rules made thereunder;

  iii. The Air (Prevention and Control of Pollution) Act, 1981 and the rules made thereunder;

  iv. The Public Liability Insurance Act, 1991.

- **Procedure:** The Courts need not follow the procedure prescribed under the Code of Civil Procedure, 1908, or the rules of evidence contained in the Indian Evidence Act, 1872. However, the proposed Court shall have all powers of the civil court including the power to punish for contempt. The Court can pass all kinds of orders, final or interlocutory. It can also award damages, compensation and grant injunctions.
- ***Locus Standi:*** The *locus standi* in original jurisdiction shall be the same as it is before High Court/Supreme Court in the writ jurisdiction in environmental matters.
- **Rule-making Power:** The environment court must be empowered to frame the schemes relating to environmental issues.

- **Repeals:** The National Environment Tribunal Act, 1995 and The National Environmental Appellate Authority Act, 1997 may be repealed.
- **Appeals:** Appeal against the orders of the proposed environment court, shall lie to the Supreme Court on the question of fact and law.
- **Power of High Courts and Supreme Court:** The powers of High Courts under Articles 226, 227 of the Constitution of India and of the Supreme Court under Article 32 of the Constitution of India shall not be ousted.
- **Principles:** The environment court is to apply the environmental principles to the technically advanced environmental issues. By application of these principles, the Courts would speed up the processes of adjudicating the environmental issues, either by imposing the pecuniary liability on the polluter, or preventing the polluter or restricting him from causing harm to the environment and improving it further for future generations. The environmental principles to be applied by the courts are:
  - **Polluter Pays Principle:** The 'polluter pays' principle was first adopted internationally, in the year 1972, at OECD Council Recommendation on Guiding Principles concerning the International Aspects of Environmental Policies. This principle of compensating the victim as well as the environment, is already laid down in Section 3(1) of the National Environmental Tribunal Act, 1995. Hence, the same can be taken up in the new proposed Act.
  - **Precautionary Principle:** This originated in the mid-1980s, under the jurisprudence of the German Vorsorgeprinzip. The decisions adopted by the States within the North Sea Ministerial Conference mark the first use of this principle in international law.
  - **The Principle of Prevention:** This principle takes care of reckless polluters who would continue polluting the environment, as paying for pollution is a small fraction of the benefits they earn from their harmful acts or emissions. Prevention of pollution must, therefore, take priority over compelling the polluter to pay the penalty.

- **Principle of New Burden of Proof:** The UN General Assembly Resolution of 1982 on the World Charter for Nature established this principle. The principle states that the activities which are likely to pose a significant risk to the nature shall be preceded by an exhaustive examination; and their proponents shall demonstrate that the expected benefits outweigh potential damage to nature. The EC's new list on hazardous wastes mentions 200 categories of wastes. Environmental Impact Assessment is one attempt intended to reduce the uncertainties attached to the potential impacts of a project.
- **Sustainable Development:** Economic development of the nation and eradication of poverty must be balanced against the conservation of the environment. Principle 3 of the Earth Summit Declaration (1992) at Rio, states that the "right to development must be fulfilled, so as to equitably meet developmental and environmental needs of present and future generations"[23].
- **Public Trust Doctrine:** The 'public trust' doctrine was referred to by the Indian Supreme Court in *M C Mehta vs. Kamal Nath*[24] case. The doctrine extends to natural resources such as rivers, forests, sea shores, air, etc. for the purpose of protecting the ecosystem. Hence, the State, as the trustee of natural resources, cannot commit the breach of trust by permitting pollution.
- **Inter-generational Equity:** The 1972 Stockholm declaration states that "man bears the solemn responsibility to protect and improve the environment for the present and future generations"[25]. Hence, the national resources of the earth must be safeguarded for the benefit of the present and future generations through careful planning and management[26].

## Constitutional Validity of Environmental Courts in India

The proposal for the constitution of environmental courts must be dealt with care, as these courts are of great importance for the millions of people all over the country. The Indian Constitution contains several provisions which require the State to either directly or indirectly protect the environment. In particular, after the Stockholm Declaration of 1972, Indian Constitution was amended[27] by incorporating schemes for protection of the environment. Hence, meeting the Constitutional goal of access to justice in matters relating to environment, is an essential facet of Article 21 of the Constitution of India. The provisions are:

- *Article 21:* Protection of life and personal liberty, extending to Right to Clean Environment.
- *Article 48A – Directive Principle of State Policy:* Protection and improvement of environment and safeguarding of forests and wildlife.
- *Article 51 A(g) – Fundamental Duty:* "It shall be the duty of every citizen to protect and improve the natural environment including forests, lakes, rivers and wildlife and to have compassion for living creatures."
- Forest and protection of wild animals and birds were brought into the Concurrent List as entries 17A and 17B in Schedule VII.

In addition to the above provisions, the Supreme Court of India too has made immense contribution to the growth of environmental jurisprudence of the country. It has entertained considerable genuine Public Interest Litigation cases under Article 32 of the Constitution; so have the High Courts under Article 226 of the Constitution. These courts have issued various directions on a number of issues concerning the environment, to the extent that they have widened the scope of Article 21 of the Constitution of India by including the right to a healthy environment. However, a number of important questions challenging the Constitutional validity for constitution of such courts were raised before the Law Commission[28] for its consideration, which were as follows:

- Can the Parliament enact a law under Article 253 of the Constitution for constituting environmental courts both in the States as well as the Center, without the State making any legislation or passing any resolutions under Article 252, especially when the matters relate to entries in the List II of the VII Schedule to the Constitution?
- Can the jurisdiction of the High Courts/Supreme Court be ousted?
- Should the jurisdiction of the normal civil and criminal courts be ousted or continued?

- Is there a need for the Environmental Appellate Court at Delhi to exercise its jurisdiction over the State Environmental Courts, or there must be a direct appeal to the Supreme Court from the State Environmental Courts?
- Should these courts follow the regular court procedure or a separate procedure?
- Should the National Environment Tribunal Act, 1995 and the National Environment Appellate Authority Act, 1997 be repealed or continued?

These questions were answered appropriately by the Law Commission, by interpreting various Constitutional provisions, and interpreting judicial precedents. One such important provision is Article 253, which provides not withstanding conditions in the chapter on legislative relations between the Union and the States. The Parliament shall have the sole power to make law to implement any international treaty. Further, the provision is in conformity with the objective declared by Article 51(c), i.e., fostering respect for international law and treaty obligations, treaty making and implementation, etc. Thus, an enactment made under Entry 13, List I, Schedule VII read with Article 253 to implement an international agreement would override and prevail over any inconsistent State enactment. Thus, the central law providing for environmental courts in India can be considered as constitutionally valid.

## Conclusion

It must be acknowledged that the credit for developing the *corpus* of environmental jurisprudence in India must go to the Supreme Court and High Courts, as well as the Parliament and State Governments. Public-spirited individuals and non-governmental organizations have risen to the occasion to highlight the problems of pollution before the judiciary and legislations, and also in enhancing public awareness. The Law Commission of India, in its 186th report, made a comprehensive study of the legal and constitutional issues involved in providing for the hierarchy of judicial mechanism in cases relating to the environment. It is hoped that the collective effort of the people and Constitutional functionaries in India will go a long way in meeting the challenges of the future.

*(Lakshmi Lella is a Faculty Member, The Icfai University (Academic Wing), Hyderabad. She can be reached at lakshmivempalli@yahoo.co.in).*

## Endnotes

1 September 2003.

2 1986 (2) SCC 176, at p. 202: "In as much as environment cases involve assessment of scientific data, it was desirable to set up environmental courts on a regional basis with a professional judge and two experts, keeping in view the expertise required for such adjudication. There should be an appeal to the Supreme Court from the decision of the environment court."

3 1996(3)SCC 212 (at p. 252): "Environmental Courts having civil and criminal jurisdiction must be established to deal with the environmental issues in a speedy manner. We may also state that the National Environmental Appellate Authority constituted under the National Environmental Appellate Authority Act, 1997, for the limited purpose of providing a forum to review the administrative decisions on Environment Impact Assessment, had very little work.

It appears that since the year 2000, no Judicial Member has been appointed. So far as the National Environmental Tribunal Act, 1995 is concerned, the legislation has yet to be notified despite the expiry of eight years. Since it was enacted by Parliament, the Tribunal under the Act is yet to be constituted. Thus, these two Tribunals are non-functional and remain only on paper."

4 1999(2)SCC 718.

5 2001(2)SCC62.

6 Prof. Malcolm Grant of Cambridge headed the "Environment Court Project". The Report on the project was published on February 18, 2000. The project was initiated and supported by the Department of Environment Transport and Regions (IETR) and had the support of the Financial Management and Performance Review of the Planning Inspectorate made in 1995-96. The purpose of the project was to study the concept of an Environment Court in the background of experience in other countries, such as Australia and New Zealand.

7 Royal Commission's 23rd Report on Environmental Planning (March 2002) (para 5.35).

8 Lord Woolf, in his Garner lecture to U K E L A on the theme, "Are the Judiciary Environmentally Myopic". The speech of Lord Woolf is printed in 1992 Journal of Environmental Law, Vol. 4, No. 1.

9 (1993) 113 S.ct. 2786.

10 AIR 1987 SC 990.

11 AIR 1988 SC 953.

12 1999(2) SCC 718.

13 2001(2) SCC 62.

14 Sec. 91 of the Code of Civil Procedure Code, 1908—Suit for public nuisance. Chapter XIV of the IPC.

16 Section 3 of The National Environmental Appellate Authority Act, 1997.

17 Section 4 of The National Environmental Appellate Authority Act, 1997.

18 Section 10(3)(b) and 10(4) of The National Environment Tribunal Act.

19 Section 12.

20 Section 38.

21 Water (Prevention and Control of Pollution) Act, 1974 and Air (Prevention and Control of Pollution) Act, 1981.

22 AP Pollution Control Board vs. M V Nayudu: 2001(2) SCC 62.

23 Bruntland Report, 1987.

24 1997 (1) SCC 388.

25 Principle 1.

26 Principle 2.

27 42nd Amendment to the Indian Constitution.

28 186th Report of Law Commissions on Environmental Courts.

# APPENDICES

# APPENDIX 1

## UN's Environment Protection Initiatives

The United Nations' initiative in combating environmental pollution is spearheaded by the United Nations Environment Programme (UNEP). UNEP acts as a catalyst in forming public opinion based on the views of the global governments and other activists and speeding up the process of environmental protection.

UNEP's mission is "to provide leadership and encourage partnership in caring for the environment by inspiring, informing, and enabling nations and people to improve their quality of life without compromising that of future generations".[1]

UNEP is headquartered in Nairobi, Kenya. It is one of only two UN programmes headquartered in the developing world (the other is UNEP's sister agency UN-HABITAT, which is also located in Nairobi). This has ensured that UNEP understands the plight of the populations affected by global pollution on a first-come-first basis.

Major Initiatives/Milestones by UNEP[2]

1975 – The first Mediterranean Action Plan UNEP-brokered Regional Seas agreement

1979 – Bonn Convention on Migratory Species

1985 – Vienna Convention for the Protection of the Ozone Layer

1987 – Montreal Protocol on substances that deplete the Ozone Layer

1988 – Intergovernmental Panel on Climate Change (IPCC)

1989 – Basel Convention on the Transboundary Movement of Hazardous Wastes

---

1 *http://www.unep.org/Documents.Multilingual/Default.asp?DocumentID=493&ArticleID=5391&l=en*

2 *http://www.unep.org/Documents.multilingual/Default.asp?DocumentID=287*

1992 – UN Conference on Environment and Development (Earth Summit) publishes Agenda 21, a blueprint for sustainable development

1992 – Convention on Biological Diversity

1995 – Global Programme of Action (GPA) launched to protect marine environment from land-based sources of pollution

1997 – Nairobi Declaration redefines and strengthens UNEP's role and mandate

1998 – Rotterdam Convention on Prior Informed Consent

2000 – Cartagena Protocol on Biosafety adopted to address issue of genetically modified organisms

2000 – MalmÃ Declaration – first Global Ministerial Forum on the Environment calls for strengthened international environmental governance

2000 – Millennium Declaration – environmental sustainability included as one of eight Millennium Development Goals

2001 – Stockholm Convention on Persistent Organic Pollutants (POPs)

2002 – World Summit on Sustainable Development

2004 – Bali Strategic Plan for Technology Support and Capacity Building

2005 – Millennium Ecosystem Assessment highlights the importance of ecosystems to human well-being, and the extent of ecosystem decline

2005 – World Summit outcome document highlights key role of environment in sustainable development

# APPENDIX 2

## Environmental Court Judgments on Tiruppur and Delhi

### Tiruppur

Tiruppur's textile industry uses bleaching liquids, soda ash, caustic soda, sulphuric acid, hydrochloric acid, sodium peroxide, and various dyes and chemicals for its dyeing and bleaching processes. This has taken its toll on the surrounding lands. Cultivation has become a hard task as high saline-sodium nature waste discharged in the Noyyal river has hardened the irrigation water. Apart from this, about 300 tonnes of colouring agents are discharged each year, into Noyyal river and other water bodies thus turning the underground water in the river basins brackish and harder.

Finally, pressure from various quarters of public and judiciary led the Tamilnadu Pollution Control Board (TNPCB) to take action for effluent treatment facilities in Tiruppur. As a result some Effluent Treatment Plants were built. But these were found to be grossly inadequate for tackling the problem.

But with no end to pollution, the Madras High Court, on Jul 14 2005, ordered the closure of more than 600 dyeing and bleaching units that form a crucial link in the hosiery industry set up. The court is maintaining constant pressure on the state government since then and has pulled up the TNPCB for not implementing its orders even though the court granted relief for the dyeing and bleaching units to restart their work on the their agreement to install zero effluent treatment plants.

But now, the industry is opposed to the court judgment on the ground of 'polluter pays' principle as it claims that the foreign exchange earned by the industry and the employment generated by them should be taken into account and the governments should subsidize the cost involved in setting up of treatment plants.

In the meanwhile, the State Government is thinking on floating a Special Purpose Vehicle (SPV) to solve the pollution problem in Tiruppur. The solution would be to lay pipelines to carry the treated effluents to the sea. Since the distance is large, the project is estimated to be around Rs.700 crores approximately. As per the government proposal, the State Government is to share 15 per cent (Rs.100 crore) of the total cost and the Union Government 25 per cent (Rs.125 crore) with rest being met by the industry. The State government is also planning to approach the High court in this regard.

If this proposal comes through, it may settle the pollution problem of Tiruppur amicably.

### Delhi

The involvement of Supreme Court in Delhi's air pollution problem stared way back in 1985 when M C Mehta, a noted environmental lawyer, filed a PIL in the Supreme Court charging that the Indian government is shirking its responsibility in controlling air pollution in the capital.

After some initial hiatus, the case assumed importance, and in 1994, the Court mandated phasing out of lead from all fuels in the four metros of the country. In 1996, it ordered that all the buses in Delhi to be converted to CNG by March 31, 2006.

But due to severe opposition from various quarters including the Government there was almost no progress in the implementation of the order. The various reasons cited were: lack of availability of CNG, high cost, need for large infrastructure etc. The Indian Government appointed the Mashelkar committee to study the issue in 2001 and this committee recommended that new emission norms to be established over the existing ones and the decision on fuel to be left to consumers. The Supreme Court rejected the committee recommendations and asked Environment Pollution (Prevention and Control).

Authority (commonly referred to as the Bhure Lal Committee) was asked to look into the matter. With its findings, the Supreme Court in its order on April 5, 2002, reaffirmed its earlier order on CNG and asked the Delhi Government to implement its orders without any further delay. The Court also advised the Indian Government to give priority to the project.

The persistence of the Supreme Court and its orders in clear terms has led to a dramatic change in the air pollution levels in Delhi. Today, Delhi is having the largest all-CNG public bus transport fleet in the world with over 10,000 buses. There are also over 90,000 vehicles of public transport in Delhi running on CNG and some 125 CNG fuel stations to refuel them.

This has contributed to reduce air pollution problems in Delhi which ranks among the the most polluted cities in the world.

# INDEX

## P



## R



## S



## T



## U



## W